*Agricultural Change in*

*Tropical Africa*

## Previous Publications

*Kenneth R. M. Anthony and Victor C. Uchendu*
  *Agricultural Change in Geita District, Tanzania*
  *Agricultural Change in Kisii District, Kenya*
  *Agricultural Change in Teso District, Uganda*

*Bruce F. Johnston*
  *The Staple Food Economies of Western Tropical Africa*
  (with Peter Kilby) *Agriculture and Structural*
    *Transformation: Economic Strategies in*
    *Late-Developing Countries*

*William O. Jones*
  *Manioc in Africa*
  *Marketing Staple Food Crops in Tropical Africa*

*Victor C. Uchendu*
  *The Ibo of Southeast Nigeria*

# Agricultural Change in
# Tropical Africa

Kenneth R. M. Anthony
Bruce F. Johnston
William O. Jones
Victor C. Uchendu

Cornell University Press   Ithaca and London

First published 1979 by Cornell University Press.
Published in the United Kingdom by Cornell University Press Ltd.,
2–4 Brook Street, London W1Y 1AA.

International Standard Book Number 0-8014-1159-9
Library of Congress Catalog Card Number 78-58039
Printed in the United States of America
*Librarians: Library of Congress cataloging information appears on the last page of the book.*

# Contents

# Tables

# Charts

# Map

# Preface

In the early days of the New Deal, when great new public works were commonplace, President Franklin D. Roosevelt's "fireside chats" provided a popular format for impersonators. One such performance that I heard had the President saying: "And furthermore I have asked the Congress to authorize construction of a new bridge over the Mississippi River—not just *over* the Mississippi River, but *all* over the Mississippi River." This book is about agricultural change in tropical Africa, but it is not all about agricultural change nor about all of Africa. Its subject matter reflects our own biases and concerns, deriving in large part from the varied experiences and previous interests of each of us. The areas of tropical Africa that receive the most attention are those we have had most opportunity to study. In large part our emphasis reflects the division between English-speaking and French-speaking Africa that was one of the unfortunate consequences of colonial rule.

The book grows out of a multidisciplinary investigation of certain aspects of African agriculture initiated in 1966, but its genesis is much earlier than that. Bruce Johnston and I had been interested in African agricultural development for some time, and we were becoming increasingly aware of our own inadequacies when trying to understand why certain innovations succeeded and others failed. We were completely disenchanted with explanations of failure that leaned heavily on "peasant conservatism" or "African communalism." While we recognized that there were social and cultural differences between African and Western societies, we ourselves did not have the expertise to judge how these differences might affect farmers' economic responses. We needed the help of someone who did. Our limited knowledge of technical aspects of farming in the African tropics was similarly embarrassing. We were aware, for example, that fertilizers were being used effectively on some African soils and that their use could be regarded as a feasible alternative to shifting cultivation in assuring soil fertility, but we lacked the technical competence to compare the potential increases in yields that might result from the use of fertilizers with increments achievable by other kinds of changes in farming practices. We needed to have working with us someone who had high competence in agricultural science combined with practical working knowledge of farming under African conditions. In fact, we were convinced that we could tackle a host of problems more effectively if we could pool our economic skills with those of a social anthropologist and an agricultural scientist who shared our interest in tropical Africa and its development.

Thanks to a grant from the Rockefeller Foundation, the Food Research Institute of Stanford University was able to appoint to its staff for a period of three and one-half years a social anthropologist in the person of Victor C. Uchendu. With this help from the Foundation we were also able to persuade the Cotton Research Corporation of London to let Kenneth R. M. Anthony, who had had many years of experience in Africa as agronomist and plant breeder, work with us on a part-time basis for the same period.

From the time Anthony and Uchendu joined us, the research effort was joint. Early in our collaboration, however, we agreed that the dominant disciplinary interest should be economic. The questions should be economic ones, even though the answers might come from sociology or plant science or farm management. Operationally, this meant that Uchendu's deep interest in land tenure might be pursued as a part of our joint effort only when tenure rules could be shown significantly to inhibit or facilitate economic response. Similar guidelines affected our following through on Anthony's concern about agricultural systems.

Interdisciplinary cooperation is not easy, but it can be rewarding. In this instance the temperament of our new partners and the strong interest each of us had in agricultural change and in Africa made it relatively simple for us to overcome problems of communication that have wrecked so many such ventures.

In writing this book we have spoken on topics that seemed important and about which we felt we had learned enough to have something worth saying. Some topics that could figure in considerations of agricultural change in tropical Africa are missing entirely, and others have received only passing mention. We did no work directed at agricultural climatology, for example, and found ourselves with nothing useful to say about the episodic droughts in the western Sudan or the incidence and coincidence of crop failure in the countries of eastern and south-central Africa. We examine certain aspects of small-farmer behavior in detail, but say little about mechanisms for coping with risk by avoidance, transfer, or distribution. Some other important matters that have been discussed by one or another of us elsewhere are given much less attention here. For example, analysis of imperfections in the agricultural marketing system and measures for its improvement are the subject of my 1972 book, *Marketing Staple Food Crops in Tropical Africa*. In addition, agricultural change and employment and interrelationships between agricultural and industrial development are treated at length in Johnston's book with Peter Kilby, *Agriculture and Structural Transformation*.

We started with a number of questions, set forth in some detail in Chapter 1, that seemed appropriate to pursue. All have policy implications and all require the insight of each of our three general disciplines. On some of these subjects we eventually felt we could speak with confidence, and they form the substance of this book: the role of the market in African agricultural change; the variety of African response to economic change; the interaction of the social structure with changes in the economic order; and the quality of the agricultural research base and of communications between research stations and farmers.

Problems of collaboration in book writing over great distances and the fact that work on the book had to take its turn with other responsibilities of each of us have delayed its completion. Uchendu served as director of the Makerere Institute of Social Research after he left the Food Research Institute in 1969 and later assumed the directorship of African Studies at the University of Illinois. Anthony had supervisory responsibility for the Cotton Research Corporation's field operation in Africa, the Middle East, and Latin America throughout the period.

In successive drafts of the manuscript, we have tried to assess the extent to which our conclusions have been overtaken by subsequent developments. Our overwhelming impression is that the factual accounts, analysis, and conclusions remain applicable, and continuing population growth has reinforced the need for emphasizing gradual but widespread modernization of labor-intensive smallholder farming. In fact, there is much wider acceptance of that view than when we adopted it as the framework for our research.

We look forward with some impatience to the time when economic advance and increased knowledge about African farming will make this book obsolete. But that time has not yet arrived.

We agreed to designate a drafting author and a rewrite author for each chapter, although things did not always work out according to plan. I was assigned overall responsibility for preparing the manuscript for the press after texts of the individual chapters had been agreed on, and I was allowed considerable latitude in rewriting. It is inevitable, I think, that the nature of our several experiences and of our collaboration should have led to a certain amount of pontificating. For this I apologize on behalf of all four authors. If the reader finds our *obiter dicta* distressing, however, he may blame those in Chapters 3 and 5 principally on me, those in Chapters 2, 7, and 8 principally on Johnston, those in Chapters 4 and 9 principally on Anthony, and those in Chapter 6 principally on Uchendu. Chapters 1 and 10 are group efforts, although the words are mostly mine, and each of us accepts responsibility for all the chapters.

WILLIAM O. JONES

*Stanford, California*

# Acknowledgments

A great many people have aided us in our investigations and in the writing of this book. Other publications growing out of our research have provoked comments that have stimulated and guided us in this attempt to marshal our findings. Each of us has benefited, too, from the assistance of his colleagues, and for this we are all appreciative. It would be vain to attempt to list all those who have opened doors and led us to new knowledge. We must, however, recognize the special help we received from Walter Falcon and Scott Pearson in supporting us through successive painful revisions and for not only showing us where we were in trouble but frequently showing us how to get out of it. We are also grateful to Robert Bates for his thoughtful comments on the manuscript.

Our field studies would not have been possible without the gracious support of members of the faculty and staff at Makerere University, the University of Ghana, and Ahmadu Bello University. To them and to the many government officers in the six countries where these studies were conducted, we express our most sincere appreciation. We earnestly hope that our research findings may partially repay the many members of ministries of agriculture, particularly their field staffs, for their conscientious support of our investigation.

Minnie Jurow carried the main burden of preparing this manuscript for publication and helped the four collaborators in many other ways throughout the long duration of this joint project. Others also contributed, most notably Helen Hoff; and we were aided in locating and analyzing materials by Rosamund Peirce, Raphael Igwebuike, and Hendrik Molster. Charles Milford merits special mention for his help with the African collections of the Food Research Institute library and for his unflagging diligence and courtesy in supplying the documentary material we required.

We are heavily indebted to the Rockefeller Foundation for its initial and continuing support, to the Cotton Research Corporation for permitting Anthony to participate, and to the Food Research Institute of Stanford University for providing the kind of intellectual climate and the support of library, statistical, and research staff that made this venture possible.

Finally we express our gratitude to the staff of Cornell University Press for their wise and patient guidance in making our manuscript a book.

Kenneth R. M. Anthony
Bruce F. Johnston
William O. Jones
Victor C. Uchendu

*Agricultural Change in*

*Tropical Africa*

# 1. Introduction

The economic advancement of the people of tropical Africa and the ability of new African nations to feed their increasing populations, and to feed them better, depend to a large measure on the improvement of African agriculture. Agricultural exports are the principal source of foreign exchange throughout much of the region; agricultural production for export and for domestic consumption is the principal contributor to national income; and millions of farms provide the principal employment. Improvement of productivity in this sector of the economy is critical for efforts in all of the other sectors.

The importance of increased agricultural production for improved nutrition is, of course, obvious for villagers who obtain most of their diet from foodstuffs they grow themselves; it may be even more critical for people who depend increasingly on purchased food. The provision of nutritionally adequate and humanly satisfying diets is essentially an economic matter. Diets are most often poor because the people who consume them are poor, and they can only be improved if consumers' incomes permit them a greater command over food resources. The ability to consume more food, more varied food, or more nutritious food may be brought about by an increase in total consumer purchasing power, however achieved, that permits the purchase of imported foodstuffs; it can also be achieved by an increase in the productivity of domestic agriculture that will bring to market a larger supply of customary foodstuffs at lower cost. The latter alternative is likely to be advantageous, however, because enlarging home-grown supplies to feed the growing urban and rural populations will reduce the competition between imports for consumption and imports that strengthen productive capacity, thus making possible more rapid economic growth, or more rapid improvement of diet, or both.

As a result of a notable decline in death rates, particularly during the past two decades, rates of natural increase in African countries have risen sharply. The rate of population growth is currently close to 2.5 percent, and it is expected to continue to rise for some time because death rates are declining with virtually no reduction in birthrates (United Nations 1971:10–11). The upsurge in the rate of natural increase has resulted in a sharp increase in the rate of growth of the population of working age. Employment in manufacturing and other nonfarm activities is still so small that even if it increases rapidly, the agricultural labor force is certain to grow for several decades at least. Agriculture affects such a large part of the population and is such a dominant component of economic activity in the tropical African countries that changes in it are bound to affect the wider processes of economic development and social modernization.

Consideration of ways in which the expansion of agricultural output and productivity can be assisted, within the context of the total economy, is a matter of first importance.

## African Farmers

Farming is carried out mainly by millions of smallholders who make their own managerial decisions; and the skill and diligence with which they apply themselves to the task of farming are the immediate determinants of the level of agricultural output and its growth. Output levels actually achieved, however, will also depend on factors beyond the control of the individual farmer which influence both his production possibilities and his incentives. The interactions between the factors operating at the farm level and socially determined conditioning factors are complex. Research and agricultural extension programs aimed at making more productive technologies available to thousands of farmers operating under diverse conditions face major problems, and there is bound to be uncertainty as to how farmers will respond to technical innovations and to price incentives.

In past years it was commonly alleged that traditional farmers would respond perversely or feebly to economic incentives. Hence, the argument ran, traditional agriculture should be largely bypassed as an unreliable, inelastic source of increased supply. At present, however, there seems to be fairly general recognition that this contention was based on incomplete understanding of the context in which traditional farmers made their decisions. The field surveys carried out in connection with this study provide further support for the view that African farmers are, by and large, rational in their behavior and in their response to economic opportunities.[1]

The beginning of the process of modern economic growth in these coun-

[1]The early statement by Jones (1960) has been enlarged upon in general terms by Schultz (1964) and confirmed by a number of econometric investigations. Especially pertinent to tropical Africa are the studies by Bateman (1965 and 1969), Dean (1965), Welsch (1964), and

tries is very recent.[2] A remarkable expansion of agricultural exports during the past half-century has been the principal stimulus to the process of transforming collections of isolated communities that were engaged predominantly in production to satisfy their own needs into national market-oriented economies. Substantial progress has been achieved in creating a network of transportation, communications, and marketing institutions to link together previously isolated communities. Nonetheless, the change in economic structure has been distinctly limited. The bulk of the population and labor force is still rural and engaged in agricultural production (if only for their own consumption), although village crafts, trading, or other activities are often carried on concurrently.

A varying but substantial fraction of agricultural output is now produced for sale. As would be expected, a large part of this commercialized output is destined for export markets because the nonfarm population dependent on purchased food remains small. Apart from some exceptions noted later, food purchases by the farm population are limited.

## Technical Change

The notable advances during the past century in perfecting experimental methods and in scientific understanding of soil science, plant nutrition, genetics, plant breeding, and other branches of agricultural science have thus far had relatively little impact on farm practices in Africa. Although enormous resources have been devoted to agricultural research in the developed countries of the temperate regions, agricultural research in tropical Africa has been limited and, until recently, confined mainly to the export crops. The special problems involved in the management of tropical soils and the associated problems of high insolation and torrential but erratic rainfall have received relatively little attention. Among food crops, some advances have been made in developing high-yielding varieties of maize that are suited to conditions in parts of tropical Africa, and programs for

---

Massell and Johnson (1968). These and other studies of the responsiveness of African farmers to prices have been reviewed by Helleiner (1972). In his study of *Agricultural Change and Peasant Choice in a Thai Village*, Michael Moerman observed that "it is common, yet quite mistaken, to think that peasant 'decision-making and technology are traditional and arational.' Men, whether in Iowa or Ifugao, farm not by instinct, but by consciously using culturally appropriate means to cope with culturally recognized goals" (Moerman 1968:27).

[2]"Modern economic growth" in Simon Kuznets's sense refers to the economic epoch, which dates from the mid-eighteenth century, during which sustained economic growth has resulted from exploiting the potential provided by science-based technology. Although the definition emphasizes the role of new technologies based on modern science, it needs to be emphasized that "application of science via technology would not have taken place without changes in social institutions" (Kuznets 1966:12).

their development and introduction in Kenya and Zambia indicate the potential contribution that well-designed research can make to increasing farm output.

The principal change to date in African farming systems has been the spread of production of a variety of export crops; apart from learning how to grow the new crops, farmers have changed the methods of production very little. For the most part, soil fertility is maintained by some system of bush fallowing, although in many areas fallow periods have been reduced considerably and a substantial number of farmers are now engaged in more or less continuous cultivation. Traditional systems of land tenure in most areas were geared to a land surplus economy based on bush fallowing. But the planting of perennial crops and other economic forces have led to important changes in the content of existing tenure systems; and in some areas, notably Kenya, the individualization of land tenure has been formalized by legal procedures for registering land titles.

Use of purchased inputs obtained from outside the agricultural sector remains distinctly limited. This result is to a considerable extent a reflection of the very low level of farm cash income. It is also related to high transport and distribution costs that make such inputs expensive. And in some areas the increased use of fertilizers is limited because research has not developed crop varieties with the genetic capacity to give a strong response to heavy fertilization. In spite of these problems, however, a growing number of farmers are now buying fertilizers, insecticides, plows, and other implements, and some are buying tractors or hiring them.

## Improving Agricultural Productivity

The very impressive achievements of the tropical African countries in the expansion of agricultural exports are reviewed in Chapter 2. Much of this growth occurred in the earlier decades of the nineteenth century and was primarily the result of private commercial efforts, facilitated by government-supported improvement of the conditions of internal travel. Concerted governmental attempts to develop the economies of tropical African countries date essentially from the end of World War II. There then ensued a great variety of agricultural development schemes in the Belgian, British, French, and Portuguese colonies, protectorates, and trust territories. This activity was continued by the new independent governments and is still a feature of the African economic landscape. On the whole, the achievements of these government efforts to develop agriculture have been disappointing. A considerable part of their failure must be attributed to the ignorance and misconceptions about the nature of African farming.[3]

The design of effective strategies for agricultural development requires an understanding of the present agricultural system, its potentialities and its constraints. There seems no doubt that farmers will respond to attractive new opportunities if they are able to. The problem is to identify the circumstances that make it possible for them to do so. Guy Hunter (1973:236) puts the matter of agricultural innovation in a nutshell: "There are three inexorable laws of innovation: the new practice must be feasible for the small farmer; it must pay him better than his present practice; and its product must be marketable. If any one of these three conditions is not fulfilled, the development of small farming systems has not failed; it has not been effectively tested." Hunter's three conditions of feasibility, profitability, and marketability are of course closely interrelated, and they transcend the customary bounds of economics to involve sociocultural and technical considerations as well. Feasibility itself, which seems purely a technical matter at first, ultimately involves considerations of costs and benefits and of cultural value systems. The question is usually not whether a farmer is physically able to adopt a new practice, but rather whether the overcoming of obstacles to its adoption will be less onerous than going without the resulting benefit. If commercial fertilizer is not available in a local market, a farmer might still be able to obtain it if he could or would meet the cost of journeying to a larger market or to a port city. Farmers have done such things, and they have obtained the means to do them by borrowing from family, friends, or moneylenders.

A lot has been learned about African agriculture in the last two decades. If more effective strategies for agricultural development are to be designed and implemented, however, it will be necessary to examine a set of questions about the nature of African farmers' problems that derive from Hunter's "inexorable laws." These questions mostly fall into Hunter's categories of feasibility and profitability, and each has technical, sociocultural, and economic aspects. In some sense they are all subsidiary parts of the grand question: Are there technologies presently available for tropical African agriculture that satisfy the criteria of feasibility, profitability, and marketability? The more detailed questions follow. Later chapters present evidence that bears variously on them.

[3]For a critique of various development schemes, see de Wilde et al. (1967) and Uma Lele (1975).

1. Technical Questions

a. Is there a potential for increasing farm output and productivity by relying mainly on innovations that make relatively small demands on scarce resources and that are consistent with an agriculture based on small farms and employing a large part of the population?

b. Are highly productive new techniques known?

c. Are methods recommended by the extension service profitable for farmers?

d. To what extent is the resource base limiting?

e. How is productivity affected by farm size?

f. How great is the scope for increasing productivity simply by propagating methods that are now practiced by Africa's more successful farmers?

2. Sociocultural Questions

a. Do existing social institutions inhibit farmers' response to market opportunities?

b. How well informed are farmers? How do they learn about new methods of production and new market opportunities? Is ignorance a major barrier to progress?

c. What have been the relative short-term and long-term effects on African agricultural production of programs relying on compulsion, exhortation, demonstration, price and income incentives, and improved marketing and transport?

3. Economic Questions

a. Is the adoption of new methods inhibited by imperfections in the markets for land, labor, and money?

b. To what extent can domestic savings and investment be relied on to provide the capital required by an improved agricultural technology? What types of capital can best be created using existing resources?

c. Is the marketing of farm tools and agricultural chemicals sufficiently reliable and economical to permit the use of these inputs?

d. If farmers increase their production of food crops, can they sell them? Is effective demand truncated by inadequate marketing systems and a strong propensity to self-sufficiency in staple foodstuffs?[4]

4. Design of Innovations

a. Is it possible to identify elements in the existing sociocultural system (e.g., communalism) and in the existing technology (e.g., mound cultivation, use of manures) whose presence or absence predisposes farmers to particular kinds of innovations?

b. What are the relative advantages of innovations requiring that only one detail in their operations be modified (single-trait innovations) compared to those that require a set of new practices (package or multiple-

---

[4]W. Arthur Lewis said of Nigeria in 1967 that "the right way to attack the stagnation of the food producing sector is to increase the demand for food. Efforts to increase supply faster than demand will come to naught" (Lewis 1967:31).

trait innovations)? Is it possible to identify particular kinds of innovative sequences that are most likely to be successful?

A number of studies made since World War II have increased our knowledge of the present character of African agriculture, and a variety of development schemes and other attempts to alter the existing organization have illuminated the problems and also the possibilities for improvement. They demonstrate that prescriptions for the improvement of African agriculture cannot be prepared from economic understanding alone but require a compound of economic, technical, and sociocultural analysis. The range of choice is bounded by technical knowledge about the productivity of various combinations of inputs, while the rate and directions of change are determined by the ability and willingness of African farmers and farming communities to alter existing patterns of production. Economists who have been asked to assist African states in the design and implementation of programs intended to foster agricultural development, and others who have attempted to formulate general hypotheses about the conditions for increased agricultural productivity in tropical Africa, have been made keenly aware of the need for specialist assistance in determining the circumstances under which specific innovations will be accepted, rejected, or modified, and in identifying and evaluating the technical possibilities for increasing farm output and productivity.

In 1965 the Food Research Institute initiated two studies of agricultural development in tropical Africa, one directed specifically at the marketing of foodstuffs and the other at the economic, sociocultural, and technical determinants of agricultural change. The two studies were closely interrelated in objectives, design, and personnel, and each drew heavily on earlier studies of African agriculture, supplemented by intensive field studies.[5] The marketing study was carried out by five teams, four led by agricultural economists and the fifth by a specialist in communications. Three of the investigators were teamed with agricultural economists from the countries where the studies were conducted. The agricultural change project was carried out by an interdisciplinary team consisting of two economists, an anthropologist, and an agricultural specialist.[6] The findings of the market-

[5]Earlier studies include accounts of agricultural change between 1945 and 1960 in various African countries that had been commissioned by the Food Research Institute. Eight of these have been published: Brown (1968); Drachoussoff (1965); Fuggles-Couchman (1965); Johnson (1964); Kettlewell (1965); Leurquin (1963); Makings (1966); and Masefield (1962).
[6]Johnston, Jones, Uchendu, and Anthony, respectively.

ing studies are reported in six monographs, a general book, and various articles. Results of the agricultural change studies have so far appeared in seven monographs and several articles.[7] The present volume undertakes to provide a general synthesis of the findings of that project.

## The Field Studies of Agricultural Change

The field studies of agricultural change were sited in areas in which significant increases in productivity had occurred and also in areas where there had been sustained governmental effort without much change. Good historical information was available for all seven areas so that the team could obtain significant information after a brief period in the field, usually six weeks. Intensive studies were undertaken in seven English-speaking countries, three in East Africa, three in West Africa, and one in central Africa, and less intensive investigations were conducted in various other countries. The studies were aimed at assessing the prevailing agricultural situation in the light of the historical record, and were intended to identify the principal determinants of change, both stimulants and deterrents. Their value for comparative purposes is enhanced because they were carried out by the same team, asking similar questions, over the comparatively short period of nineteen months.[8]

The studies sought to determine the technical and economic possibilities for increasing farm productivity, the nature of the agricultural extension effort, the farmers' awareness of innovations that were available through the extension service or private merchants, and the circumstances under which farmers were apt to be more or less willing to undertake an innovation. The explanation of receptivity was sought in the profitability of particular innovations, the availability of land, capital, and credit, the nature of the traditional farming system, and the adequacy of marketing facilities. Factors deriving from the social system were watched for as well. The investigations undertook to determine the influence on agricultural change of governmental intervention in agricultural marketing, specifically the establishment of parastatal marketing boards and marketing cooperatives,

[7]The marketing study monographs and book are: Alvis and Temu (1968); Gilbert (1969); Jones (1972); Mutti and Atere-Roberts (1968); Thodey (1968 and 1969); and Whitney (1968). Agricultural change monographs are: Anthony and Johnston (1968); Anthony and Uchendu (1970, 1974); Uchendu and Anthony (1969, 1975a, and 1975b); and Uchendu (1969).

[8]The studies in Uganda, Kenya, Tanzania, Zambia, and Bawku District (Ghana), were carried out by Anthony and Uchendu, that in Northern Nigeria by Anthony and Johnston, and that in Akim-Abuakwa (Ghana) by Uchendu alone.

the subsidization of farm credit and purchased inputs, and the provision of transportation and communication facilities. Attention was also directed to the effectiveness of private enterprise in providing markets for products and supplies of inputs, and to the impact on small farmers of large farms and plantations, whether public or private.

Each study began with visits to research stations in order to learn what technical means were available for increasing productivity—what agents of change could offer to farmers. Research staff communities have greater continuity of service in an area than do extension staff, and are therefore likely to be better informed about local obstacles to the adoption of new techniques.

After the visit to the research stations, information was sought about the means used to inform farmers of new crops and methods of cultivation and about the activities of supporting institutions—the size and training of extension staff, staff training facilities, the extension approach, marketing facilities for products and inputs, activities of cooperative societies, availability of production credit, and the nature and extent of subsidies.

Most of the time, however, was devoted to interviewing farmers. A total sample of sixty was drawn in each area except Kisii, where it was decided to draw samples in two localities that were distinct in agricultural potential and achievement. Emphasis was on obtaining a good general knowledge of the farmers' problems and of their attitude to change. Where the concept of "improved" or "progressive" farmer existed, the sample was made up of fifteen representative progressive farmers suggested by the local agricultural department staff and forty-five of their neighbors who were selected by the investigators. Elsewhere, farmers were selected from lists of registered growers (Akim-Abuakwa), members of cooperative societies (Geita), or on the advice of village heads (Katsina). In Bawku, the Agricultural Officer provided the names of ten farmers and the team selected five more in each of four villages.

Questionnaires following a common pattern but appropriate to each area were prepared after the preliminary conversations with research and extension staff and a reconnaissance tour of the area. Information was collected about the number of persons supported by the farming unit, the amount of time spent on off-farm activities, the education of members of the farm family, the crops grown and their relative importance, the size of the farm, the number of separate parcels of land per farm, and how each was acquired. Areas were determined by pacing and measurement. (Farmers rarely had any concept of unit area.) Information was obtained about cash receipts, expenditures on inputs like seed, fertilizers, insecticides, contract plowing, hired labor, and feedstuffs, and the nature and amount of live-stock and implements owned by the farm operator. Farmers were also asked about the location of markets for their products and their sources of credit. Farm interviews were especially directed at determining the extent to which agricultural department recommendations were understood and prac-ticed and farmers' reasons for following or ignoring them. All interviewing was done by one of the three investigators, and usually went beyond the questions listed in the set schedule. Only rarely were farmers reluctant to answer questions.

The results of these studies are reported in detail in the seven mono-graphs referred to in n. 7 above. A synthesis is presented in Chapter 5, and various findings are referred to in other chapters.

## Some Biases and Limitations of This Study

It should be made clear at the outset that this study has been affected by certain predispositions of the authors. The general approach has been con-ditioned by their beliefs that human nature is much the same everywhere, that the structural transformation of the African economies will require many decades, and that general statements about the nature of African societies are possible and useful. The study has also been influenced by the authors' long experience with the fragility of African data and by their strong bent for learning as much as possible about the nature of an econ-omy before prescribing remedies for its ills. The reader may benefit from a fuller explanation of this warning.

### *Human Behavior*

This collaborative effort began with the conviction that African farmers are motivated by the same kinds of desires and concerns as are most people, including the desire to improve their state of physical well-being. We do not assume that African farmers put maximization of their economic returns above all other goals, nor do we suggest that though farmers in Africa are poor they are also efficient. The individual who is completely dominated by economic motives is rare in any society, including African ones. Economic drive varies from individual to individual; there is no reason to believe that its general level or variability is greater or smaller in Africa than elsewhere.

For the purposes of economic analysis it is not even necessary to assume that economic motivation is as dominant in the societies of tropical Africa as in those of the Western world. The relationships postulated by economics will hold if the desire for economic betterment is present to some degree in a great many individuals and if they are able to order their affairs to serve this objective.

## Evolution of African Economies

The African states have barely begun to undergo the process of structural transformation that will lead to the explosion of production in the nonagricultural sectors characteristic of modern economic growth. This process can be expected to require many decades, and until it is rather well advanced, most African families will find their principal support in farming. At present most African farms are small, family operated, and employ moderately labor-intensive methods. In most parts of Africa, despite the availability of uncultivated land, any rapid move toward labor-displacing methods would almost certainly result in an increase in farm size and widespread unemployment and displacement of populations. Largely for this reason, our concern has been almost exclusively with small farms, and there is little in this book about group farms, state farms, or cooperatives.

In considering the place of agriculture in the economies of tropical Africa (Chapter 2), some attention is given to the interrelationships between agricultural development and overall economic growth. Research initiated during the past few years should lead to improved understanding of interactions in both product and factor markets between agriculture and the nonfarm sectors. Moreover, it is now generally recognized that simple two-sector models are inadequate and that it is essential to consider the distinctive characteristics of (a) the urban large-scale or "formal sector," (b) the urban small-scale or "informal sector," and (c) the rural nonfarm sector, and the relationships among those sectors and agriculture (Byerlee and Eicher 1974; ILO 1972). The pattern of agricultural development will, of course, exert a strong influence on the nature of those intersectoral relationships. Moreover, the experience of countries where the process of structural transformation is further advanced suggests that the stimulus to the growth of nonfarm output and employment is likely to be especially strong when a country's agricultural strategy fosters the progressive modernization of a large and growing percentage of the small-scale farm units, which inevitably predominate when the bulk of a country's population depends on agriculture for work and income (Johnston and Kilby 1975). The focus of this study, however, is on the process of agricultural change itself and on the factors that permit widespread increases in farm productivity and income.

## The Homogeneity of African Societies

African societies are extremely diverse in their social and political organizations and in their value systems. There are also important differences in the economic base of populations living in the rain forest and the savanna, in uplands and lowlands. But the economies also have much in common, especially within certain broad ecological zones. In most places, land is plentiful relative to labor and both are plentiful relative to capital. Most families rely on their own production of food crops or animal products for a large part of their food calories. Farmers still rely almost entirely on human energy for their field and barnyard operations, even for transport. Because of these considerations, and because land is relatively plentiful, farmers are almost entirely free of the kinds of landlord-tenant constraints that characterize the other continents and North Africa. In their decision-making they are much like owner-operators nearly everywhere. Tenure is secure as long as the land is occupied, a most important consideration for farm-management decisions. Restriction of the rights of alienation and inheritance limits optimum combination of land and management and helps to exacerbate fragmentation, but it is not of direct importance in day-to-day farm operations.

Similarities among regions and countries that grow out of the economic base are reinforced by the fact that during the first half of this century all but two of the tropical African countries were under the sovereignty of European powers. As a result, agricultural strategies were frequently more influenced by the consensus of minds at the metropolitan centers than by the heterogeneity of affairs in the subject area.

Because farming activities, development problems, and development possibilities tend to be more or less the same over large geographical regions like the savanna zone of the Sudan, the rain forest of the Congo and the Guinea Coast, and the savanna plateaus of eastern and southern Africa, we are encouraged to attempt to formulate propositions that will hold true over wide areas. If some of these propositions appear to be equally applicable to other populations in other continents, the reader should not be too surprised.

## Sources of Information

This study tends to focus on areas where economic change of some significance has taken place, recently or in the recent past. We believe that it is more informative to examine possible generators of change than it is to review all of the possible inhibitors, and that the causes of change are more likely to be observable in places where change has occurred. In addition to this historical bias, the selection of countries from which information was obtained has been influenced by the accessibility of particular African countries to research activities, the availability of published and documentary information about them, and by the prior experiences of the authors. The concentration on English-speaking countries has already been noted. Selection of supportive material is eclectic and illustrative, except for the case studies.

## The Nature of the Findings

The agricultural change project from which this book comes was conceived as a search for tentative answers to some questions about the nature of African agricultural systems that were believed to be critical for development planning. It was an exploratory study that derived from the growing body of knowledge about African economies. No attempt was made to achieve the rigor required by formal hypothesis testing because of the very unsatisfactory character of most national data, because of the character of the questions, and because of the cost of generating data that would be satisfactory.

The book, too, is exploratory and cannot pretend to be definitive. It is primarily concerned with things as they are and have been, not as they might or should be. It attempts to be more factual than prescriptive, more positive than normative. But it was undertaken in order to achieve normative ends. Its purpose is to provide a firmer base for designing plans to change the shape of African agricultural economies.

# 2. The Place of Agriculture in
# the Economies of Tropical Africa

The predominance of agriculture in employment and in national output, which is characteristic of less developed countries, is especially conspicuous in the countries of tropical Africa. The considerable concentration of economic activity in farm households located in relatively isolated and largely self-sufficient rural communities reflects the primacy of man's need for food. And this predominantly agrarian structure, characteristic of preindustrial societies with low levels of productivity, has not yet been substantially modified. The initiation of modern economic growth has been too recent and its impact too restricted to have resulted in significant structural transformation.

In the mid-1960s, approximately 80 percent of the economically active population in the countries of tropical Africa were engaged in agriculture. Except for the implausibly low estimates for Benin and Congo, only in Ghana did the share fall below 60 percent (see Table 2.1).[1] On the other hand, agriculture accounted for just over 40 percent of total Gross Domestic Product (GDP). The intercountry variation in agriculture's share in GDP is rather large, reflecting above all large variations in the importance of mining. Thus in Gabon and Liberia agriculture represented just over one-

[1]The usual problems that characterize statistical data for African countries are compounded by uncertainties and inconsistencies related to the definition of economically active population that is used. It is likely, for example, that the census figure for Ghana showing only 58 percent of the economically active population in agriculture is below the true figure because of substantial underreporting of the number of women in the farm labor force in northern Ghana. FAO estimates for Nigeria (1976a:35) indicate that the percentage of the economically active population in agriculture declined from 71 percent in 1960 to 62 percent in 1970. Those figures are derived from labor force estimates which indicate that agriculture absorbed only about one-fifth of the 22 percent increase in the total economically active population between 1960 and 1970. The figures also imply that the nonfarm labor force increased 58 percent between 1960 and 1970. Such a rapid increase is conceivable, given the large expansion of the army and the impact of the civil war on the economy, but it would be very unusual in relation to the experience of other developing countries (Johnston and Kilby 1975, esp. 83–85). For the entire continent, the recent FAO estimates suggest a 5 percent reduction from 77 to 72 percent between 1960 and 1970; but those figures are influenced by the relatively low share of agriculture in Egypt and other North African countries and also in South Africa.

*Table 2.1.* Proportionate share of agriculture in employment (1970), gross domestic product (1965–73), and total exports (1974) in selected tropical African countries*

| Country | Percent of economically active population in agriculture 1970 | Agriculture as percent of GDP 1965–73 | Agriculture as percent of total exports 1974 |
|---|---|---|---|
| Angola | 64 | ... | 46[a] |
| Benin | 50 | 39 | 91[b] |
| Burundi | 87 | 65 | 99 |
| Cameroon | 85 | 33 | 72 |
| Central Africa | 91 | 34 | 44 |
| Chad | 90 | 50 | 97 |
| Congo | 42 | 17 | 13 |
| Ethiopia | 84 | 56 | 89 |
| Gabon | 82 | 18 | 1 |
| Gambia | 82 | 58 | 94 |
| Ghana | 58 | 44 | 76 |
| Guinea | 85 | 49 | ... |
| Ivory Coast | 84 | 29 | 65 |
| Kenya | 82 | 34 | 55 |
| Liberia | 76 | 24 | 20 |
| Madagascar | 89 | 30 | 74 |
| Malawi | 89 | 51 | 90 |
| Mali | 91 | 45 | 65 |
| Mauritania | 87 | 34 | 13[b] |
| Mozambique | 74 | ... | 74 |
| Niger | 93 | 50 | 37 |
| Nigeria | 62 | 46 | 6 |
| Rhodesia | 64 | 17 | ... |
| Rwanda | 93 | 67 | 71[a] |
| Senegal | 80 | 35 | 44 |
| Sierra Leone | 72 | 33 | 21 |
| Somalia | 85 | ... | 94 |
| Sudan | 82 | 39 | 97 |
| Tanzania | 86 | 42 | 72 |
| Togo | 73 | 44 | 20 |
| Uganda | 86 | 53 | 92 |
| Upper Volta | 87 | 46 | 92[b] |
| Zaïre | 80 | 19 | 13 |
| Zambia | 73 | 8 | 2 |

*Data for Column 1 from FAO, *Production Yearbook 1975* (Rome, 1976a); Column 2 from IBRD, *World Tables 1976* (Washington, 1976); Column 3 from FAO, *Trade Yearbook 1975* (Rome 1976b), and United Nations, *Yearbook of International Trade Statistics 1975,* Vol. 1 (New York, 1976a).
[a]1973.
[b]1972.

quarter of total GDP, whereas mining accounted for nearly 20 percent of the total in Gabon and slightly over 30 percent of GDP in Liberia. In Zambia, where mining represented close to 40 percent of GDP, agriculture's share was only 10 percent.

Notable variations in the importance of minerals as a commodity export also explain substantial differences in the share of agricultural products in total exports. Thus agriculture accounted for a mere 3 percent of total exports in Zambia, where copper dominates the economy, an even smaller share in Gabon, where minerals and forest products are the major exports, and less than 25 percent in five other mineral-exporting countries. As a weighted average for thirty-two countries, agriculture represented 55 percent of total commodity exports. That average is, however, heavily weighted by the value of exports in the mineral-exporting countries; in half of the countries, agriculture's share exceeded 70 percent (Table 2.1).

During the period since 1965 there has been a slow decline in the share of agriculture in total GDP; and in three countries which have recently become important exporters of mineral products—Nigeria, Senegal, and Togo—the share of agriculture in exports declined drastically.[2]

The first section of this chapter presents a brief description of the major food crops and dietary patterns in the countries of tropical Africa. Attention is also given to the nature of nutritional problems in the region and the influence of rapid population growth on the food supply situation. The discussion of commercial production of agricultural commodities in the second section focuses mainly on the factors that have influenced the growth of agricultural exports and presents a summary view of the long-term trends in export expansion. For the period since 1950, attention is also given to the way in which foreign exchange earnings from agricultural exports have been influenced by price fluctuations as well as changes in quantities exported. The next sections deal respectively with agricultural processing and with farm-supplied and purchased inputs. In the final section some of the interrelationships between agricultural expansion and overall economic growth are noted. That important but complex topic is not pursued very far, however, because the emphasis of this book is on changes *within* the agricultural sector.

## Characteristics of the Food Economies of Tropical Africa

Although there has been a notable expansion of commercial production since the turn of the century, the food economies of tropical Africa are still

[2]See Table 2.5 below.

heavily oriented toward subsistence production. Only about half of total agricultural output is marketed, and export crops account for somewhat more than half of the share that is commercialized. There are, however, notable variations between and within regions in the importance of commercial production, and some examples of that variability are examined shortly. There are also rather striking variations in the composition of local diets, reflecting regional variations in food production.[3] Starchy staple foods are the dominant sources of calories and other nutrients throughout tropical Africa (apart from some relatively small pastoral populations that rely heavily on milk, meat, and sometimes animal blood). But the particular commodities that are of major importance show striking regional variations, and in places there is also considerable seasonal variation in the importance of different staple foods.

In the more humid regions, one of a half-dozen starchy roots, tubers, or banana-plantains is the major staple food, whereas in the drier grassland areas, millets, sorghums, or maize are typically the dominant staples. The spread of cassava (manioc) and maize, two crops introduced from the New World, have considerably modified African diet patterns during the past three centuries. Particularly during the past half-century, cassava has frequently replaced yams or some other traditional crop as the dominant staple and maize has supplanted sorghum in many areas.[4] But the two crops are also important secondary staples in both forest and savanna regions and are the most widely distributed of the African staple food crops. Although rice cultivation has also spread during the past half-century, it is a major crop only in the traditional rice areas—Madagascar and along the coast of West Africa from the Casamance River in southern Senegal to the Bandama in the center of the Ivory Coast, and in a few districts of central and eastern Africa.

The dominance of starchy staple foods is a well-nigh universal characteristic of food consumption in low-income countries, but the percentage of calorie intake provided by one or more starchy staples in some African diets is exceptionally high. Even in urban areas, from 60 to 85 percent of the calorie intake usually derives from cereals, roots or tubers, or banana-plantains, and in rural areas where cereals are the dominant staple the percentage may be 80 percent or more. In spite of this heavy reliance on the cheaper sources of food energy and nutrients, expenditures for food and the imputed value of subsistence production represent a large fraction of total

---

[3]Variations in the position of major food crops are, of course, closely correlated with the agroclimatic regions that are described in Chapter 4.

[4]For a somewhat more detailed description of African diet patterns and references to the literature, see Jones (1972:chap. 2).

family income. For urban households, outlays for food typically account for 50 to 70 percent of total household expenditure.[5] Diets of urban consumers also show substantial regional variation, reflecting growing conditions in nearby producing areas. Specialized production and long-distance trade in staple foods are still the exception. More expensive products such as meat, fish, cowpeas, and kola nuts figure more prominently in trade. Rural as well as urban households frequently rely on purchases of meat, dried fish, and some of the other ingredients of the sauces which accompany the main dish (usually a stiff porridge or paste) and provide variety as well as high-quality protein and other nutrients to supplement the starchy staple foods in the diet.

Evidence from food-consumption surveys and other sources has tended to indicate that undernutrition—deficiency of calories—is of only limited importance in tropical Africa, apart from special circumstances such as the Nigerian war in the late 1960s and the 1928–29 and 1943 famines in Ruanda-Urundi. The 1943 famine was caused in part by a serious drought, but the crushing blow was a serious attack of the same fungus (*Phytophthora infestans*) that was mainly responsible for the Irish potato famines of the nineteenth century; the high-altitude zones of Rwanda and Burundi are among the few localities in tropical Africa where white potatoes have become an important staple food. Less serious famines have been a periodic problem in the savanna regions where rainfall in a normal year is only marginally adequate for crop production, and a succession of dry years gave rise to very serious food problems in the Sahel region in the early 1970s. The Sahelian drought, which began in 1968 and was intensified by a succession of years with below-normal rainfall (1970–1973), resulted in severe undernutrition and hardship and an unknown but considerable increase in human mortality. Pastoral areas to the north of the savanna farming zone suffered the sharpest reduction in rainfall, and estimates of stock losses range from a 20 to 50 percent death rate among animals (Caldwell 1975:17–27). In earlier years, locust attacks often produced famine conditions in these areas, but the extensive spread of cassava

[5]The percentage fell within the 50 to 70 percent range in all but one of twenty-five budget surveys carried out in nineteen cities in tropical Africa during the 1950s (Kaneda and Johnston 1961:235). Individuals employed by governments and by large private firms obtained substantial wage increases during the 1960s, so the percentage of expenditure devoted to food has probably declined. But for rural households and for workers in traditional manufacturing and service activities, it has probably changed very little. Even a majority of those in the relatively high-wage modern sector probably continue to devote a high fraction of their total household expenditure to food because of supporting relatives seeking urban jobs, or working only part-time as casual laborers, or eking out a meager income in the services sector.

as an antifamine crop has substantially reduced the incidence of famine from that cause. Cassava is not vulnerable to locust attacks; and the roots can be held for a considerable period in the ground, to be harvested as needed. Colonial administrators frequently resorted to compulsion as well as persuasion to spread the practice of planting an acre or two of cassava as a famine-reserve crop. International measures to control locust invasions have also achieved considerable success.

As noted earlier, the scattered evidence available suggests that malnutrition has been a more serious problem in tropical Africa than undernutrition. However, for reasons noted shortly, undernutrition appears to be emerging as a more serious problem in a number of countries. Because of heavy reliance on starchy roots and banana-plantains with their exceptionally low protein content per thousand calories, there appears to be a rather high incidence of protein deficiency among infants in the postweaning stage and other groups that require a relatively high percentage of protein calories in their diet. Severe quantitative and qualitative deficiencies in protein intake, often aggravated by gastrointestinal infections that cause a reduction in the utilization of protein and also reduce appetites, give rise to kwashiorkor, a serious nutritional syndrome that is often fatal if not treated in time. Deficiencies of protein and of B vitamins are much less of a problem in the savanna regions, where diets are based on cereals, but other nutritional deficiencies may be serious. Vitamin A deficiencies, leading in severe cases to blindness resulting from xerophthalmia, are particularly likely to appear at the end of a long dry season because of a lack of fresh vegetables and fruits that are good sources of vitamin A and other vitamins and minerals. Vitamin deficiencies are also likely to be a problem in the cities, where fruits and vegetables tend to be expensive and many of the greens and other wild products that are sources of nutrients in the countryside are not available.

The nutritional value of diets, which is the relevant consideration, obviously depends not only on the composition of the starchy staples but on the nutrients provided by the other foods that round out the diet. Even very small quantities of meat and fish are important in that regard, and beans and other sources of good-quality vegetable protein are in general even more important in offsetting the quantitative and qualitative deficiencies of the proteins obtained from the dominant staple foods. In some areas, green leaves (including cassava leaves) are also significant sources of high-quality protein and of the B vitamins. Rising incomes and wider knowledge of the nutritional requirements for a good diet will tend to lead to improvements in the nutritional quality as well as the economic value of African

diets, whereas urbanization per se may have unfavorable nutritional consequences, at least in the short run. With growing awareness on the part of
governments of the significance of good nutrition for health, well-being,
and productivity, programs to achieve more rapid improvements in nutritional status are likely to receive increased attention.

The characteristics of African diets that have been briefly noted suggest
that measures aimed at improving the quantity and quality of protein or the
intake of vitamin A are likely to be of particular importance depending on
the region. For consumers who depend on purchased food, various programs of enrichment or fortification of products such as *gari* (cassava
meal) offer a relatively low-cost and practical means of upgrading diets.
Promoting the manufacture and widespread distribution of products of high
nutritional value to meet special needs represents an additional possibility
for improving the nutritional status of vulnerable groups. These possibilities are closely related to the development of domestic food-
processing industries, and are discussed in a later section of this chapter
that examines the growth of agricultural processing. For the large fraction
of the population which relies predominantly on home-produced foods, the
major need is to develop recipes for weaning foods that can be prepared in
the home by combining locally available foods to produce nutritionally
balanced mixtures suitable for infant feeding.

Food production statistics for the countries of tropical Africa are subject
to large margins of error.[6] Estimates of year-to-year changes in food
supplies need to be treated with great caution, but nevertheless there are
disturbing indications that the expansion of food production has failed to
keep pace with the growth of population in a number of countries in
tropical Africa. This is suggested very crudely by FAO's regional indices
of per capita food production. Table 2.2 compares the estimates for the
African continent with estimates for North and Central America, South
America, and Asia for the five-year period ending in 1976. The exceptionally low figure for 1973 was influenced considerably by the sharp reduction
in food production in Senegal, Mali, Chad, and other countries affected by
the Sahelian drought. However, the possibility that more fundamental
trend factors may be operating is suggested by the fact that, according to
the FAO estimates of index numbers of food production, the average
annual rate of increase declined from 2.6 percent for the 1961–70 decade to
only 1.5 percent for the period 1970–76.

[6]A paper by D. G. R. Belshaw (1975) contains a good general discussion of the problems of
estimating crop production of smallholders in tropical Africa, a detailed analysis of the
deficiencies of the various crop estimates available for Uganda, and a useful bibliography.

*Table 2.2.* Estimated food production per capita for Africa and other continents, 1972–76 (1961–65 = 100)*

|                              | 1972 | 1973 | 1974 | 1975 | 1976 |
|------------------------------|------|------|------|------|------|
| Africa                       | 99   | 92   | 98   | 96   | 97   |
| North and Central America    | 106  | 107  | 107  | 112  | 114  |
| South America                | 101  | 101  | 104  | 103  | 111  |
| Asia                         | 103  | 106  | 105  | 109  | 109  |

*Data from FAO, "Special Feature: FAO Indices of Agricultural Production," *Monthly Bulletin of Agricultural Economics and Statistics* 26 (1977), p. 26.

In past decades, expansion of food production in tropical Africa has been primarily a result of a "horizontal" increase in output based on increased inputs of land and labor. The population growth which led to increased demand also provided the additional labor required to bring additional land under cultivation. Much of this "horizontal" expansion of food production has probably been achieved with little or no decline in the marginal productivity of labor. Typically, the new land has not been much inferior to the land already in cultivation; and when expansion has become possible because of eliminating barriers such as tsetse infestation and trypanosomiasis, the land thus brought into cultivation has sometimes been of higher quality. But as a result of the increase in cultivated area resulting from expansion of production for sale and the acceleration in the rate of population growth since the 1940s, the scope for expanding the area under cultivation has been much reduced. In many areas increasing population pressure has led to a shortening of the fallow periods between crop cultivation, with a consequent reduction in soil fertility, more difficult problems of weed control, and often a deterioration in soil structure and erosion.[7] On the basis of a careful survey of the literature, William Hance has estimated that nearly 40 percent of the land area in sub-Saharan Africa was "subject to pressure" by the late 1960s (Hance 1970:421). The adverse effects on food supplies would have been much more serious and visible if it had not been for the scope that existed for shifting to crops, notably cassava, that give a tolerable yield even on quite infertile soils.

[7]Relying on a number of surveys of farming systems in various parts of Ghana, George Benneh estimates that in the bush-fallow system, the most widespread farming system in the country, the fallow period has been shortened in recent years from six to ten years to two to three (Benneh 1973:139). In Kenya, rapid growth of population in "high potential" areas has led to substantial migration to "medium potential" areas characterized by marginal and erratic rainfall. This has resulted in exceedingly rapid growth of the farm population in those marginal areas that has significantly increased the magnitude of famine problems in years when rainfall is below normal (Lynam 1977).

There is, of course, great variation among and within countries in the extent to which population growth is giving rise to food-supply problems. Even though the available evidence is subject to a considerable margin of error, it seems clear that rapid population growth has aggravated problems of food supply and accentuated the need for technical innovations that will permit widespread increases in agricultural productivity and output.

## Commercial Production of Agricultural Commodities

Since the turn of the century, and especially since World War II, there has been a tremendous spread of commercial production of agricultural commodities. Initially commercial production was geared almost entirely to export markets, and even today export production is the dominant source of farm cash income in many African countries.

### Factors Influencing the Growth of Exports

By and large the expansion of agricultural exports in tropical Africa appears to have been of the "vent-for-surplus" variety, in which a country having large quantities of labor or natural resources available at low opportunity costs can increase output rapidly by employing these factors to produce commodities for export.[8] In the classification of vent-for-surplus models employed by R. E. Caves, the experience of most African countries is best described by the "surplus resources" version rather than by the "unlimited labor" version (Caves 1965). Despite Africa's prominent role as an exporter of slave labor to the Americas in the eighteenth and nineteenth centuries, the African countries themselves experienced shortage of labor relative to resources once it became profitable to export minerals and products of the soil.

J. S. Hogendorn has examined Myint's vent-for-surplus model in the light of African experience up to 1914. His analysis of peanuts in Nigeria and Senegambia, palm oil along the Bight of Benin, cotton in Nigeria and

---

[8]The term "vent-for-surplus" derives its present vogue from an article by Hla Myint in 1958 (Myint 1971:chap. 5). It was originally used disparagingly by J.S. Mill in referring to Adam Smith's statement that countries benefit from foreign trade because "it carries out that surplus part of the produce of their land and labour for which there is no demand among them. It gives a value to their superfluities..." (Smith 1937:415). Mill regarded "the vulgar theory ... that deems the advantage of commerce to reside in the exports ... a vent for its surplus" to be "a surviving relic of the Mercantile Theory." He pointed out that "the country produces an exportable article in excess of its own wants from no inherent necessity, but as the cheapest mode of supplying itself with other things" (Mill 1902:386–87). Myint changed Smith's concept to apply to surpluses of factors, not of products (Myint 1971:chap. 5). See also Jones (1969:279–83).

Uganda, kola in Ashanti, and cocoa in the Gold Coast and western Nigeria revealed "several threads of common experience" (Hogendorn 1976:27). Specifically, he emphasizes six similarities and contrasts: (1) Demand increased very rapidly and suddenly for all the commodities he studied except kola. (2) Labor, not land, constrained production, and shortages were made good by slaves and migrants. (3) In all instances, imported goods were important incentives for expanding production for export. (4) Improved transport was important for cocoa and Senegambian peanuts and essential for Uganda cotton and Nigerian peanuts, but was not a factor in the growth of kola and palm oil trade. (5) Indigenous capital formation was important. (6) Indigenous initiative was of critical importance in numerous ways. The last two findings are contrary to Myint's expectations. Sara Berry's perceptive analysis of the economics of cocoa development in Nigeria also emphasizes that the vent-for-surplus model "neglects the role of capital formation in the growth of West African agricultural exports and, in so far as it implies that expanded export production entailed little or no change in the traditional organization or structure of the rural economy, it is positively misleading" (Berry 1975:3).

Prior to the improvements in transportation and communications that enabled export-import firms to link African cultivators with the world market, the commercial demand for agricultural output was severely limited and "effective demand" consisted overwhelmingly of the "reservation demand" of rural households for their own subsistence. Given such a restricted market and inelastic demand, much of the land was left idle or used only for hunting,[9] and a relatively small amount of total available labor time was devoted to agricultural production. Introduction of a market for cocoa, cotton, peanuts, or some other "cash crop"—a term that has been virtually synonymous with "export crop" in the African literature— abruptly increased the value of the additional output that could be produced by putting more land under cultivation and increasing the amount of labor employed in farming. At the same time, the availability of attractive European substitutes for African-produced tools, utensils, and textiles facilitated transfer of labor out of these activities. With the spread of European rule, other labor was released from legislative, adjudicative, and security duties.[10]

Export-crop industries that resulted from the opening of European markets to African producers had some important features. Rapid increases in

[9]A large amount of land must have been in buffer zones separating alien polities. In a sense it, too, was employed—in maintaining the peace.

[10]Land once locked up in buffer zones was also released for other uses.

crop production were possible and could be sustained over long periods with only a minimal investment of outside resources and with little change in agricultural technology other than learning how to grow a new crop using techniques which (like the traditional food crops) employed virtually no inputs aside from the traditional ones of land and labor. Moreover, the release of labor from craft manufactures and other nonagricultural activities meant that the export production need not reduce the production of food crops for the family's needs. Thus cultivators could, as Myint said, "hedge their position completely and secure their subsistence minimum before entering into the risks of trading" and look upon the goods purchased with the cash receipts from their new crop as "a clear net gain obtainable merely for the effort of the extra labour in growing the export crop" (Myint 1971:135). The additional income from production of the new crop was not, of course, costless; the labor employed had an opportunity cost in terms of other economic production or leisure foregone. (In a later period when political and administrative changes led to involuntary unemployment, the opportunity cost may sometimes have been negative.[11]) Much of the available labor time was apparently perceived as having a low opportunity cost, and the new farming opportunities were sufficient to induce significant changes in labor's allocation.[12] But the opportunity cost of labor in African countries is not zero, and shortages of labor may limit farm output even though the farm labor force is not fully utilized according to some concepts.[13] Up to the present it seems to have been mainly new export opportunities that have led to increases in the allocation of labor to farming, but in the future new technologies or market opportunities associated with the growth of domestic demand will no doubt play a similar role in many areas.

Substantial and continuing expansion has been possible primarily because the spread of transportation links and of knowledge of the income-

[11]Igor Kopytoff said that in the 1950s underemployment of the Suku in Zaïre was so great "that even with one fourth of the men absent (from the village) those who remain suffer from essential boredom" (Jones 1968:2).

[12]On the general question of the allocation of labor time to various activities, see Jones's summary account of a conference on competing demands for the time of labor in traditional African societies sponsored by the Social Science Research Council (Jones 1968). Edgar Raynaud (1969) presents a good review of available evidence on the use of labor time in African societies and a general discussion of the reasons why the strict dichotomy between work and leisure that tends to characterize a modern industrial economy has limited applicability to the agricultural sector in developing countries. This matter is discussed further in Chapter 3.

[13]A comparative analysis of survey data for nearly fifty farming communities in ten countries shows the amount of labor time devoted to agriculture to be surprisingly low (Cleave 1974).

earning opportunities of export crops has meant that new cultivators continued to join the ranks of the earlier producers until the supply of land available for enlarged production became a limiting factor. The supply of labor available in areas well suited to high-value crops such as cocoa or coffee has typically been augmented by migration of workers from less favored regions.

There have, of course, been numerous variations in the way in which this process unfolded in different parts of tropical Africa.[14] In many areas the peak seasonal labor requirements for the export crop occurred during the slack season for traditional crops. In such areas, the main effect of the new crop enterprise was to reduce the time devoted to nonagricultural activities. Elsewhere, seasonal labor bottlenecks resulting from introducing a new crop led to a shift in the major staple food. Thus in a number of areas manioc or plantain production replaced millet, sorghum, or other staple crops because of their low labor requirements (per thousand calories) and a seasonal labor pattern that did not involve sharp conflict with the labor requirements of the export crop.

In a large part of tropical Africa the scope for increasing labor inputs in farming was substantial because of the traditional division of labor by sex, which left men a good deal of time to spend on activities that were no longer possible or needed as European influence spread, or on activities that were readily compressible when more attractive alternatives became available. In many societies women and children were responsible for most of the work of producing and preparing food crops, the farm work of men being limited to the heavy tasks of felling trees and clearing bush plots. Where export-crop production has been initiated, the traditional restrictions related to the division of labor by sex have generally been modified and the new cash crop has most often been a "man's" crop. The sharp curtailment in tribal warfare with the spread of law and order has increased the time that could be allocated to farming. The time devoted to hunting has declined as game has become less abundant. The time available for farming was also increased when porterage was replaced by transportation by rail, truck, motor-powered river craft, and bicycle. Moreover, the new imported goods that provided an incentive for export-crop production also led to a decline in traditional artisanal activities such as basketry, spinning and weaving, pottery making, and ironworking.[15]

[14]For some concrete examples and references to the literature relevant to this type of export growth, see Johnston (1964:151–55).

[15]Stephen Hymer and Stephen Resnick (1969) have especially emphasized these nonagricultural activities carried on by farm households or in small-scale service and artisan estab-

Other time-consuming tasks of an earlier period such as settling disputes, formulation of laws, and education have been taken over by the government to a considerable extent, and healing and worship have been preempted in part by specialists. During the colonial period the African population was sometimes forbidden to engage in these service activities, and in numerous instances commerce in certain kinds of merchandise was also proscribed. Acceleration in the growth of the labor force since World War II, together with the relatively slow growth of new employment opportunities outside agriculture, has also tended to keep the opportunity cost of labor low. On the other hand, there are reports which suggest that shortages of farm labor sometimes arise because awareness of the relatively high wages paid in the modern urban sector has raised the reservation price of labor above the wages paid by farm enterprises (ILO 1972:43).

Smallholders have frequently hired migrants and other farm workers, but plantations and plantation workers have been relatively unimportant in tropical Africa. Plantation production has been dominant only for rubber, sisal, and tea, and there is significant smallholder production even of these crops. In recent years smallholders have accounted for most of the expansion in tea production and a significant part of the growth of rubber production, especially in Nigeria, which emerged as the largest producer in the region in the late 1960s. However, expansion in Liberia, a plantation producer, was more rapid in the 1970s, with the result that it regained the lead by a considerable margin in production and in exports by a much larger margin (because of increased domestic consumption in Nigeria). Sugar is also produced as a plantation crop, but until fairly recently Uganda was the only country in tropical Africa with a sizable sugar industry. More important than plantation production has been the cultivation of export crops— mainly tobacco, pyrethrum, and coffee—by European farmers in Rhodesia, Kenya, Angola, and other areas of white settlement. In some instances, notably coffee, production did not differ greatly from that on plantations. For these crops, too, there was substantial expansion by African smallholders after World War II in areas such as Kenya, where previ-

---

lishments in the village. Their "partial list" of such activities includes processing of food and fuels, spinning, weaving, metalworking, dressing and tanning of leather, manufacture and repair of tools and implements, pottery and ceremonial objects, as well as investment in house building, fence repairing, and services such as recreation, protection, transport, and distribution. To this list, Jones (1969) adds such service activities as transmission of culture and education of the young, arriving at a consensus on matters of general concern, settlement of disputes by village palaver or before formal courts, transmission and creation of oral literature, sculpture and painting, entertainment and recreation through music and dance, healing of the sick, and propitiation of the supernatural.

ous restrictions were removed and positive encouragement provided. Production of oil palm presents a mixed picture. On the eve of independence, exports of palm oil from the former Belgian Congo, which came about equally from European plantations and outlying groves, had caught up with exports from Nigeria, where smallholder production is dominant.[16] As a result of increasing domestic consumption and unfavorable price policies, Nigeria's exports of palm oil had declined substantially prior to the drastic reduction that occurred during the Nigerian civil war. In recent years the most rapid expansion of oil palm acreage has been in the Ivory Coast where the Société pour le Développement et l'Exploitation du Palmier à l'Huile (SODEPALM), a quasi-governmental corporation, has undertaken plantation-type production based on high-yielding varieties that have substantially increased the profitability of oil palm production. There has also been considerable expansion of oil palm acreage in Benin and the Cameroons, carried out by similar quasi-governmental corporations.

## Expansion of Agricultural Exports

A summary picture of the expansion of agricultural exports of nine major commodities for the extended period 1909–75 is given in Chart 2.1.[17] The individual countries shown in the various panels of the chart have typically accounted for close to 90 percent of total exports from Africa of each of the commodities except cotton, and peanuts and peanut oil. Throughout most of the period since 1909 there has been an enormous expansion in exports. There has, however, been a marked leveling off of this expansion process since about 1965. Moreover, in contrast with the fairly consistent upward trend characteristic of most countries in earlier decades, there have been more pronounced intercountry variations in export performance since 1965. Domestic disturbances, increased domestic consumption, and adverse

[16]There was some plantation production of palm oil in West Africa in the precolonial period. F. E. Forbes, a British naval officer who visited Dahomey in 1850, was introduced to a palm oil plantation by an influential family of Brazilias (the term used to describe liberated slaves who returned to West Africa from Brazil). Pierre Verger (1953) includes this account in his interesting collection of texts about the Brazilias, but the following quotation is from the English original (Forbes 1851:123): "March 21st.—The Souza family having invited me to a pic-nic, and promised to show me a European plantation,—started at noon in hammocks, and, at a distance of three miles to the westward, found they had not exaggerated their description. A splendid palm-oil plantation was before me, thickly set with palm trees, intermixed with corn, cotton, yams, and cassada, according to the soil. . . . The proprietor was a liberated African from Bahia, originally a Mahee; and the plantation in the highest order."

[17]Unless otherwise noted, this discussion of agricultural export and price trends is based on the following sources: FAO (1976b, 1976c, 1976d, 1977, and earlier issues) and UN (1976c and earlier issues).

*Chart 2.1.* Agricultural exports of selected African countries, 1909–75* (thousand metric tons)

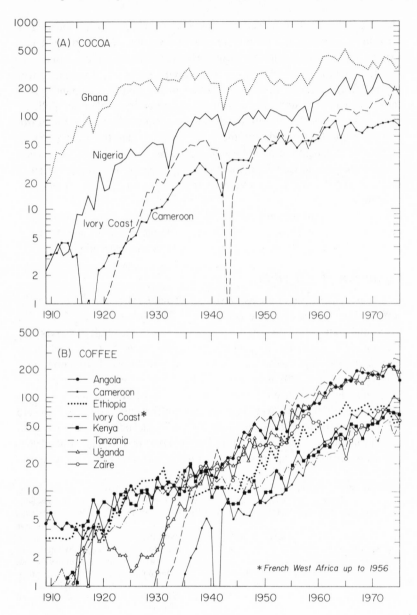

*Chart 2.1.* (cont.)

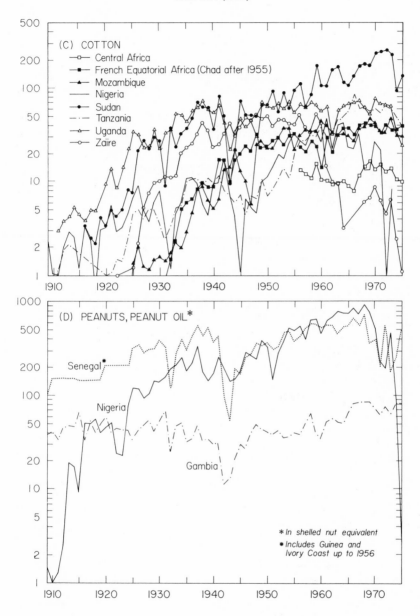

(C) COTTON
- Central Africa
- French Equatorial Africa (Chad after 1955)
- Mozambique
- Nigeria
- Sudan
- Tanzania
- Uganda
- Zaire

(D) PEANUTS, PEANUT OIL*

Senegal*

Nigeria

Gambia

*In shelled nut equivalent
*Includes Guinea and
 Ivory Coast up to 1956

*Chart 2.1.* (cont.)

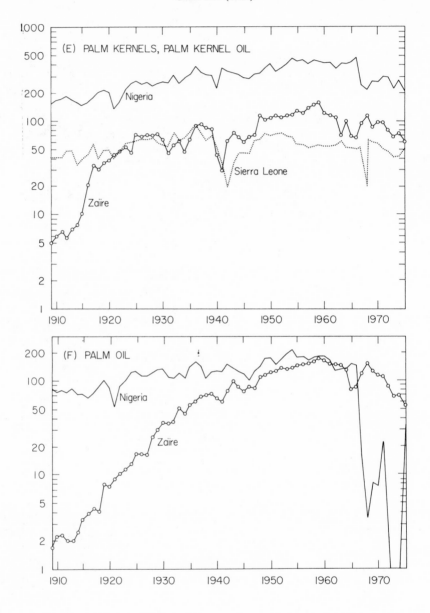

*Chart 2.1.* (cont.)

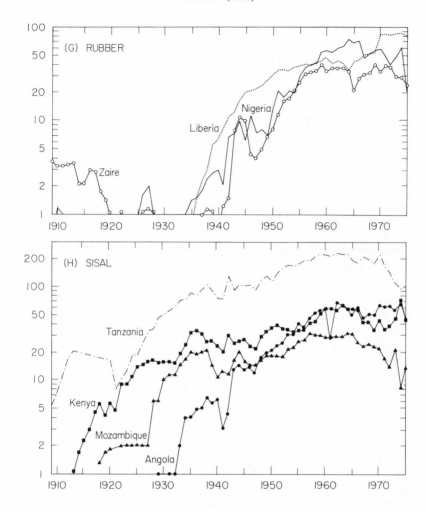

\* Data from Food Research Institute compilation, based principally on International Institute of Agriculture (IIA), *Yearbook of Agricultural Statistics,* Food and Agriculture Organization (FAO), *Yearbook of Agricultural Statistics, Pt. II, Trade,* and FAO, *Trade Yearbook,* supplemented by official statistics of foreign commerce of individual countries and by trade publications.

price policies have all been factors contributing to a marked reduction in agricultural exports in certain countries.

In the years between the end of World War II and the mid-1960s, the countries of tropical Africa were able to expand their earnings from agricultural export crops more rapidly than developing countries in other regions. The value of exports of ten major commodities in 1960, 1966, 1969, and 1975 is shown in Table 2.3, together with Africa's share in total world exports by value and by volume. (Africa's share in the world total by weight is also shown for those years and for 1934 to 1938 and 1950.) A quantity index for the ten commodities shown in Table 2.3 rose by about 60 percent during the 1950s and continued to increase at about the same rate until 1965, when it reached 200 (1950 = 100). The index then remained at about that level through 1969 and, as noted shortly, only tea has registered substantial growth since 1969.

The expansion of coffee exports during the period since World War II has been especially significant. Exports of coffee from tropical Africa rose from an annual average of 130,000 tons during the 1934–38 period to nearly a million tons in 1966. This raised Africa's share from 8 to 30 percent of world exports in terms of weight (Table 2.3). The value of exports from Africa rose from $650 million in 1966 to $1.2 billion in 1975, but this mainly reflected a rise in coffee prices. Since the early 1960s, the expansion of African exports has been only a little more rapid than the increase in total world exports. Thus a 23 percent increase between 1961–65 and 1966–70 resulted in a moderate increase in Africa's share from 28 to 31 percent. In 1974, exports from African countries were about equal to the record volume of 1973 and world exports were down substantially because of a large reduction in exports from Brazil. As a result, the continent's share of the world total was 35 percent in that year, but in 1975 it was again at 31 percent, reflecting a recovery of world exports and a small decline in African shipments.

The value of Africa's cocoa exports rose even more rapidly during the decade ending in 1975, amounting to close to $1.2 billion in that year. This reflected a nearly fourfold increase in cocoa prices from the depressed level of 1966 and put cocoa in second place as a source of foreign exchange. Although cocoa exports were higher in the early 1970s than in the late 1960s, 1972 was the only year when exports were substantially above the average level reached in the 1961–65 period. Africa's dominance of world cocoa exports reached 78 percent in 1966, but in both 1960 and 1975 the share was 71 percent.

Earnings from cotton exports rose only from $611 million to $986 million between 1966 and 1975. Thus cotton dropped to third place as a source

Table 2.3. Africa's exports of ten major commodities and share in world total*

| Year | Coffee | Cotton | Cocoa | Peanuts and peanut oil | Tobacco | Palm Oil | Palm kernels and oil | Rubber | Tea | Sisal | Total |
|---|---|---|---|---|---|---|---|---|---|---|---|
| | Value of exports from African countries (1,000 U.S. dollars) | | | | | | | | | | |
| 1975 | 1,227,142 | 986,304 | 1,189,512 | 417,354 | 233,869 | 81,746 | 90,249 | 89,765 | 138,757 | 112,522 | 4,567,211 |
| 1969 | 651,320 | 633,300 | 570,272 | 271,522 | 87,568 | 26,381 | 78,385 | 82,710 | 79,680 | 40,657 | 2,521,795 |
| 1966 | 650,480 | 610,520 | 345,067 | 331,842 | 173,940 | 52,939 | 103,190 | 77,410 | 71,402 | 61,317 | 2,478,107 |
| 1960 | 357,560 | 643,100 | 394,796 | 236,883 | 117,660 | 78,340 | 125,386 | 107,920 | 41,017 | 78,271 | 2,180,933 |
| | Africa's share in world total by value (percent) | | | | | | | | | | |
| 1975 | 29.2 | 22.3 | 72.7 | 50.8 | 9.1 | 8.3 | 49.6 | 5.6 | 13.4 | 71.0 | |
| 1969 | 26.3 | 29.4 | 73.3 | 73.9 | 6.7 | 23.0 | 76.2 | 5.0 | 14.0 | 55.4 | |
| 1966 | 27.1 | 26.4 | 75.3 | 76.8 | 14.1 | 36.9 | 86.4 | 5.9 | 11.2 | 60.6 | |
| 1960 | 18.8 | 26.0 | 72.1 | 76.0 | 12.1 | 63.3 | 87.8 | 5.9 | 6.3 | 63.1 | |
| | Africa's share in world total by weight (percent) | | | | | | | | | | |
| 1975 | 31.2 | 15.7 | 70.5 | 57.8 | 11.2 | 11.2 | 49.5 | 5.3 | 16.4 | 75.4 | |
| 1969 | 29.0 | 24.6 | 74.4 | 77.2 | 7.9 | 22.2 | 74.9 | 4.5 | 15.7 | 58.0 | |
| 1966 | 30.2 | 20.6 | 78.2 | 77.0 | 17.5 | 37.7 | 84.9 | 5.6 | 13.1 | 60.3 | |
| 1960 | 24.2 | 19.1 | 71.2 | 75.9 | 13.4 | 64.2 | 86.2 | 6.1 | 7.8 | 63.1 | |
| 1950 | 15.2 | 22.5 | 66.1 | 58.6 | 13.2 | 69.2 | 86.8 | 2.4 | 3.9 | 51.5 | |
| 1934–38 (average) | 7.9 | 18.2 | 67.0 | 34.4 | 5.7 | 52.8 | 82.0 | .7 | 1.6 | 42.1 | |

*Data from FAO, *Trade Yearbook 1975, 1970, 1967, 1966, 1953* (Rome). Countries in tropical Africa account for virtually all of the continent's exports of these commodities with the exception of cotton, for which Egypt's share is substantial.

of foreign exchange, whereas in 1960 it ranked well above coffee and cocoa as Africa's major export crop. The 80 percent increase in cotton prices was a little larger than the rise in coffee prices, but cotton exports declined from 800,000 tons in 1966 to a little over 600,000 tons in 1974 and 1975 after reaching a peak in 1971 of nearly a million tons. Not surprisingly, Africa's share in world exports by weight dropped from 21 percent in 1966 and 25 percent in 1969 to only 16 percent in 1975. Exports of peanuts and peanut oil still ranked fourth as a source of foreign exchange earnings in 1975. Although the volume was down considerably from the record levels reached in the mid-1960s, the unit price of peanuts and peanut oil considerably more than doubled between 1966 and 1975. The decline in the volume of peanut and peanut oil exports was reflected in a decline in Africa's share in world exports from 77 percent to 58 percent in 1975.

Palm oil was a significant exception to the general tendency for Africa's share in world exports of these major commodities to increase. After a small decline from 69 to 64 percent of the world total between 1950 and 1960, Africa's share dropped sharply to 22 percent in 1969. Nigeria's exports had declined considerably even before the civil war in the late 1960s and have remained at an extremely low level. Zaïre's exports have also declined substantially. Although palm oil exports from the Ivory Coast rose sharply from virtually nil in 1969 to over 110,000 tons in 1975, this was less than one-third the combined exports of Nigeria and Zaïre in 1960. Moreover, largely as a result of a huge expansion of exports from Indonesia and especially Malaysia, world shipments of palm oil rose to 2.0 million tons in 1975, nearly a threefold increase since 1966. As a result, Africa's share in total exports was down to 11 percent in 1975.

Tea is the only major crop for which Africa's share in world exports has increased significantly during the period since 1965. After rising from about 8 percent of world exports in 1960 and 13 percent in 1966, Africa's share rose to over 16 percent in 1975.

Most of the other export crops were characterized by considerable year-to-year fluctuations in the late 1960s and early 1970s, but the general tendency was one of stagnation, especially as compared with the buoyant growth in the years before 1965. African sisal exports declined a little in quantity and as a share of total world exports between the first and second half of the 1960s, and that trend has continued in the 1970s. The decline during the 1970s was particularly marked in Tanzania, the major sisal exporter. In Kenya and Mozambique the declining trend was reversed briefly in 1973 and 1974 in response to large increases in sisal prices in those years.

## Fluctuations in Export Prices and Terms of Trade

A few general observations concerning price changes during the 1950s and 1960s will help in interpreting the volume changes summarized in Chart 2.1 and the value totals for ten commodities for selected years given in Table 2.3. It is also important to pay at least brief attention to changes in the terms of trade of the countries of tropical Africa as influenced by changes in import prices as well as variations in the prices of export commodities. Increases in the price of petroleum and a number of other imported products were, of course, especially sharp in 1973 and 1974; prices of some of Africa's export crops also rose sharply in those years, whereas the increases for other crops were relatively small.

In the early 1950s all ten of the commodities shown in Table 2.3 reached peak price levels which were influenced considerably by the Korean war and by the strengthening of import demand as a result of the postwar economic recovery in European countries. The peaks for oils and oilseeds, cotton, sisal, and rubber were reached in 1951 or 1952, whereas coffee and cocoa were at their highest levels in 1954 and tea a year later. In most instances the peak levels were some 50 percent higher than average prices prevailing in 1950, but cocoa prices in 1954 were 70 to 80 percent above the 1950 level.

Export prices in 1969 were well below the high levels reached in the early 1950s, but the overall change from 1950 was surprisingly small. The price increases for a few crops such as coffee and cocoa, which rose considerably at the end of the 1960s, nearly offset the effects of price declines for other crops, with the result that a price index for the ten commodities declined only from 100 to 97. Consequently, the aggregate growth of export receipts between 1950 and 1969 was nearly the same as the twofold increase in the volume of exports.

Cocoa prices were especially volatile during the 1950s and 1960s; the average price in 1965 was less than one-third the 1954 peak. Year-to-year fluctuations were also sizable for other commodities, but the average rate of change in prices between 1951 and 1970 was rather small. Thus the annual increase over that twenty-year period averaged only 0.4 percent for coffee and 0.7 percent for cocoa and groundnut oil (but 1.5 percent for groundnuts). The other major commodities registered a decline over that same period. The average annual rate of decrease for palm oil was only 0.1 percent, and for tea the decline averaged 1.3 percent. Cotton and sisal prices declined at an average rate of 2.1 percent between 1951 and 1970; rubber registered the sharpest decline at an annual average rate of 3.2

percent. It is significant that the last three commodities all faced increasing competition from synthetic substitutes.

The first half of the 1970s was characterized by exceptionally large fluctuations in both export and import prices. The enormous increase in petroleum prices in 1973 and 1974 was a major factor responsible for a deterioration in the terms of trade of all countries apart from Nigeria and Gabon. In addition, the marked acceleration in inflation in industrialized countries in 1973 and 1974 led to substantial increases in the import prices of most manufactured products. The export prices of some of Africa's agricultural commodities have also increased sharply since the late 1960s. For example, an increase in the unit value of Ghana's cocoa exports from $55 to $75 per metric ton between 1972 and 1973 was associated with a new peak in cocoa prices which slightly exceeded the 1954 record price. Because of the rise in import prices, however, in real terms the 1973 price was substantially below the 1954 peak. Cocoa prices registered another increase of approximately 50 percent in 1974 because of a tight world supply situation resulting from continuing drought in West Africa. After a decline in 1975, cocoa prices rose again in 1976 and 1977.

Coffee prices also increased considerably in 1973 and 1974, reaching a level in 1974 some 60 percent above 1972 and 80 percent above the average level prevailing in the late 1960s. But the really sharp increases in coffee prices occurred in 1976 and especially in 1977, when New York wholesale prices rose above $3 a pound. This was mainly a consequence of severe frost damage in Brazil, but a sizable reduction in Angola's exports in 1975 and 1976 accentuated the reduction in world supplies.

The increase in sisal prices in the first half of the 1970s was very abrupt. The European import price for East African sisal doubled between 1972 and 1973 and again between 1973 and 1974, bringing sisal prices to a record level and reversing the serious slump of the late 1960s and early 1970s. However, sisal prices declined in 1975 and remained at a lower level through 1976. Recent changes in rubber prices have followed a somewhat similar pattern. After falling to a twenty-two-year low in 1972, rubber prices doubled in 1973 with a further small increase in 1974. Prices declined in 1975 but recovered in 1976. Cotton prices also doubled between 1972 and 1973, for the first time exceeding the peak levels reached in 1951. They then declined moderately, but have remained above the 1951 level.

Most of the other African export crops also registered price increases in 1973 and 1974, but at lower rates. Tea prices were essentially unchanged until 1974, when a 17 percent increase over the previous year brought tea prices to a level 12 percent above the average prevailing in the late 1960s but only 7 percent higher than the average in the 1961–65 period. Further sizable increases in 1976 and 1977 appear to have been influenced by some tendency on the part of consumers in importing countries to shift from coffee to tea.

The terms of trade of most African countries are influenced strongly by fluctuations in commodity prices because of their heavy dependence on agricultural exports as a source of foreign exchange. In the 1950s and 1960s there appears to have been a gradual deterioration in the terms of trade. This was followed by a more erratic pattern in the 1970s, when most countries experienced a sharp deterioration in their terms of trade and balance of payments position. Although the average level of commodity prices did not decline appreciably during the 1950s and 1960s, the price of manufactured imports rose at an average annual rate of 1.8 percent between 1951 and 1970, which was sufficient to bring about a considerable worsening of the terms of trade for most countries.

A marked improvement in the terms of trade in the early 1970s was cut short by the enormous increase in petroleum prices in 1973 and 1974 and an accelerated increase in the prices of imported manufactured products. The fivefold increase in petroleum prices has benefited Nigeria and Gabon, but for the remaining countries it has meant an enormous increase in the cost of imports. The worldwide increases in the unit value of manufactured imports—5 percent in 1971, 8 percent in 1972, 16 percent in 1973, and 19 percent in 1975—have also contributed to a worsening of the balance of payments position of most countries in tropical Africa. The situation has naturally been most acute in countries that rely heavily on the commodities which have not experienced very substantial price increases, and worst of all where an unfavorable composition of exports has been associated with a reduced volume of exports. Changes in a country's level of reserves as a percentage of the value of imports provides a useful summary indicator of the status of its balance of payments position. Table 2.4 presents the ratio of reserves to imports in 1970 and 1975 in Gabon and Nigeria, the two petroleum-exporting countries which have experienced a marked improvement in their exchange position, and in the remaining countries, a number of which have experienced a sharp deterioration in their reserve position.

## Intercountry Variations in the Importance of Agricultural Exports and Rates of Growth

Variations in the importance of export production are very great, although in all countries some combination of agricultural and forest prod-

*Table 2.4.* Ratio of reserves*ᵃ* to imports for petroleum-exporting countries and others, 1970 and 1975*

|  | Percent | |
|---|---|---|
|  | 1970 | 1975*ᵇ* |
| Petroleum exporters | | |
| Gabon | 18.4 | 40.5 |
| Nigeria | 21.0 | 96.4 |
| Others | | |
| Benin | 24.4 | 9.8 |
| Burundi | 68.9 | 71.1 |
| Cameroon | 33.4 | 3.4 |
| Central Africa | 4.1 | 8.0 |
| Chad | 3.7 | 3.3 |
| Congo | 15.1 | 11.9 |
| Ethiopia | 41.5 | 97.1 |
| Gambia | 45.1 | 49.2 |
| Ghana | 14.1 | 18.5 |
| Ivory Coast | 30.6 | 5.5 |
| Kenya | 49.7 | 17.0 |
| Madagascar | 21.8 | 12.6 |
| Malawi | 29.5 | 23.5 |
| Mali | 1.9 | 1.7 |
| Mauritania | 5.4 | 39.9 |
| Mauritius | 61.1 | 46.8 |
| Niger | 32.2 | 28.1 |
| Rwanda | 27.0 | 26.1 |
| Senegal | 11.3 | 7.2 |
| Sierra Leone | 33.8 | 15.9 |
| Somalia | 46.8 | 46.4 |
| Sudan | 7.5 | 3.6 |
| Tanzania | 20.4 | 7.5 |
| Togo | 54.8 | 27.5 |
| Upper Volta | 73.3 | 50.4 |
| Zaïre | 35.7 | 4.2 |
| Zambia | 93.1 | 12.6 |

*Data from United Nations, Department of Economic and Social Affairs, *U.S. Economic Surveys, 1975* (New York, 1976c), pp. 75–76.

*ᵃ*Reserves include gold, convertible foreign exchange, special drawing rights, and reserve position in the International Monetary Fund.

*ᵇ*Imports are valued CIF. Figures for Benin, Central Africa, Congo, Gabon, Madagascar, Mauritania, Somalia, Upper Volta, and Zaïre, are for 1974.

ucts and minerals dominates the export trade. It will be seen from Table 2.5 that the per capita value of exports in 1965 ranged from a mere $4 in Burundi, Rwanda, and Upper Volta to over $200 in Gabon. (As noted shortly, the range has become even greater in recent years.) At the lower end of the scale were seven countries with per capita exports in 1965 of $10 or less per year, including three of the four landlocked territories of West Africa (Chad, Niger, Mali, and Upper Volta). The export trade in those countries is almost entirely dependent on agricultural products, but their agricultural resources are poor because of low rainfall and sandy soils; and their competitive position in world markets is further impaired by the high cost of inland transport.[18] At the opposite extreme are four countries with per capita exports of approximately $100 or more. The Ivory Coast, with per capita exports of $77 in 1965, was the only country that depended primarily on agricultural exports which came close to the top four countries.

The considerable intercounty variation in the rate at which foreign exchange receipts from agricultural crops expanded during the 1950s and 1960s has, of course, been influenced by the impact of price changes as well as by different rates of change in the quantities exported. In Uganda, for example, the rise in export proceeds in the mid-1950s was especially sharp because the percentage increase in the price of the Robusta coffee that Uganda exports was even greater than the average increase for coffee. But in the late 1950s and 1960s, export receipts were affected adversely by sharp declines in the prices of both coffee and cotton, the country's two principal export crops. It has been estimated that a virtual doubling of the volume of exports between 1954–56 and 1967 resulted in an increase in aggregate farm income from these crops of not much over 10 percent (Mackenzie 1971:38). Ghana's experience was similar. An approximate doubling in the volume of cocoa exports between the mid-1950s and mid-1960s was almost exactly offset by a sharp decline in cocoa prices; and the total value of Ghana's exports in 1965 was at about the same level as in 1954, reflecting the continuing dominance of cocoa in total exports.

On the other hand, some countries in tropical Africa managed to continue to expand their foreign exchange earnings and cash income from agricultural exports in spite of lower prices for major products. Thus, agricultural export earnings in Tanzania continued to expand in the 1960s because the increase in the volume of exports and expansion of new export

[18]The drought conditions of the late 1960s and early 1970s exacerbated the economic problems of the six Sahel countries—Chad, Mali, Mauritania, Niger, Senegal, and Upper Volta. A monograph by Elliot Berg (1975) provides a very useful analysis of those effects and of the more basic economic handicaps facing those countries.

*Table 2.5*. Total population, per capita value of exports and of gross domestic product, and the share of agriculture in exports in selected tropical African countries, 1965, 1973, and 1974*

| Country | Total population (million) | | Gross domestic product per capita (U.S. dollars) | | Total exports of merchandise per capita (U.S. dollars) | | | Agricultural exports as percent of total | |
|---|---|---|---|---|---|---|---|---|---|
| | 1965 | 1974 | 1965 | 1973 | 1965 | 1973 | 1974 | 1965 | 1974 |
| Angola | 5.2 | 6.2[a] | 90[b] | 490[b] | 39 | 126 | . . . | 76 | 46[c] |
| Benin | 2.4 | 3.0 | 60[b] | 88 | 8 | 13[d] | . . . | 71 | 91[d] |
| Burundi | 3.2 | 3.7 | 45[b] | 80[b] | 4 | 8 | 8 | . . . | 99 |
| Cameroon | 5.3 | 6.3 | 111 | 187 | 27 | 57 | 76 | 73 | 72 |
| Central Africa | 1.4 | 1.8 | 129 | 143 | 24 | 22 | 27 | 42 | 44 |
| Chad | 3.3 | 4.0 | 65[b] | 64 | 11 | 13 | 14 | 92 | 97 |
| Congo | 1.1 | 1.3 | 120[b] | 284 | 47 | 52 | 58 | 6 | 13 |
| Ethiopia | 22.7 | 27.2 | 47 | 72 | 5 | 9 | 10 | 99 | 89 |
| Gabon | .5 | .5 | 369 | 863 | 228 | 735 | 1,954 | 2 | 1 |
| Gambia | .3 | .5 | 85 | 125 | 41 | 41 | 79 | 97 | 94 |
| Ghana | 7.7 | 9.6 | 265 | 234 | 38 | 61 | 68 | 73 | 76 |
| Guinea | 3.5 | 4.3 | 75[b] | 96 | 14 | . . . | . . . | . . . | . . . |
| Ivory Coast | 3.8 | 4.8 | 210[b] | 326 | 77 | 154 | 253 | 67 | 65 |
| Kenya | 9.5 | 12.9 | 86 | 149 | 23 | 38 | 48 | 57 | 55 |
| Liberia | 1.4 | 1.7 | 180[b] | 298 | 127 | 223 | 235 | 24 | 20 |
| Madagascar | 6.0 | 7.8 | 80[b] | 115 | 15 | 27 | 31 | 92 | 74 |
| Malawi | 3.9 | 4.9 | 40[b] | 85 | 10 | 21 | 25 | 92 | 90 |
| Mali | 4.6 | 5.6 | 60[b] | 63 | 7 | 11 | 11 | 96 | 65 |
| Mauritania | 1.0 | 1.3 | 114 | 68 | 55 | 98[d] | . . . | 5 | 13[d] |
| Mozambique | 7.0 | 9.0 | 65[b] | 380[b] | 15 | 26 | 32 | 77 | 74 |
| Niger | 3.5 | 4.5 | 82 | 85 | 10 | 15 | 12 | 92 | 37 |
| Nigeria | 48.7 | 61.2 | 68 | 118 | 13 | 49 | 156 | 63 | 6 |
| Rhodesia | 4.5 | 6.1 | 233 | 299 | 98 | . . . | . . . | 46 | . . . |
| Rwanda | 3.1 | 4.1 | 50[b] | 76 | 4 | 8 | 9 | 57 | 71[c] |
| Senegal | 3.5 | 4.3 | 163 | 220 | 43 | 48 | 91 | 89 | 44 |
| Sierra Leone | 2.4 | 2.7 | 136 | 146 | 35 | 46 | 49 | 15 | 21 |
| Somalia | 2.5 | 3.1 | 55[b] | 84 | 14 | 18 | 20 | 84 | 94 |
| Sudan | 13.7 | 17.3 | 96 | 104 | 15 | 24 | 25 | 99 | 97 |
| Tanzania | 11.7 | 14.8 | 69 | 99 | 17 | 26 | 28 | 79 | 72 |
| Togo | 1.7 | 2.2 | 89 | 138 | 21 | 29 | 85 | 62 | 20 |
| Uganda | 8.6 | 11.1 | 83 | 108 | 26 | 29 | 30 | 80 | 92 |
| Upper Volta | 4.9 | 5.9 | 50[b] | 61 | 4[e] | 4[d] | . . . | 81 | 92[d] |
| Zaïre | 17.6 | 24.2 | 65[b] | 113 | 24 | . . . | 60 | 12 | 13 |
| Zambia | 3.7 | 4.8 | 206 | 339 | 98 | 244 | 287 | 3 | 2 |

*Data for population from United Nations, *Demographic Yearbook 1975* (New York, 1976b); gross domestic product and gross national product from IBRD, *World Tables 1976* (Washington, 1976); gross national product from FAO, *Trade Yearbook 1970* and *1975* (Rome, 1971a and 1976b) and from United Nations, *Yearbook of International Trade and Statistics 1975*, Vol. 1 (New York, 1976a).
[a]Extrapolated from 1972.
[b]Gross national product.
[c]1973.
[d]1972.
[e]1964.

crops, such as tobacco and cashew nuts, more than offset the decline in coffee and cotton prices. However, the increase in the value of exports to $300 million in 1974, a 60 percent increase from 1969, was mainly a result of higher export prices for coffee, cotton, and sisal, which offset a reduction in volume for those major exports. The Ivory Coast is the most striking example of sustained growth in income and foreign exchange earnings from agricultural exports. The value of total exports in 1965 was two and one-half times its 1954 level. Prior to the rapid expansion of timber exports in the 1960s, cocoa and coffee dominated Ivory Coast exports, with earnings from coffee being about twice as large as the export proceeds from cocoa. The experience with cocoa paralleled the situation in Ghana, whereas export proceeds for coffee registered a modest increase because the expansion in volume more than offset the price decline. The major difference, however, was that in contrast to Ghana, exports of several other products—most notably timber but also bananas, pineapple, and palm oil and kernels—expanded rapidly. Whereas cocoa and coffee represented more than 90 percent of total exports in 1954, this percentage had declined to 48 percent in 1965 (Stryker 1974:30).

As previously noted, the intercountry variation in the per capita value of exports has become extremely large in recent years. Rapid expansion of mineral exports has been mainly responsible for this widening divergence. Estimates of per capita GDP for 1965 and 1973 and of the per capita value of exports for both 1973 and 1974 are included in Table 2.5. The most dramatic increases have occurred in the petroleum-exporting countries of Nigeria and Gabon. In Nigeria, the rise in exports was from $13 per capita in 1965 to $49 in 1973 and $156 in 1974. Because of its small population and high-value exports of iron ore as well as petroleum, the per capita value of Gabon's exports reached $735 in 1973 and $1,954 in the following year. Of the countries that have continued to rely mainly on agricultural exports, the Ivory Coast registered the largest increases in per capita value of exports: from $77 in 1965 to $154 in 1973 and $253 in 1974. Increased proceeds from cocoa and coffee exports accounted for about half of the increase in total export receipts from $458 million in 1969 to $1.2 billion in 1974. Although higher prices contributed considerably to that increase, it is to be noted that the Ivory Coast is an exception to the usual pattern revealed in Chart 2.1, in that cocoa and coffee exports continued to expand quite rapidly through 1974. The previously mentioned expansion in palm oil exports from a negligible level in 1969 to over 100,000 tons in 1974 as the SODEPALM plantations came into production was also a factor.

At the opposite extreme, in Central Africa, Chad, Congo, and Mali, there was virtually no change in the per capita value of exports in spite of the considerable increase in export prices between 1965 and 1973 and 1974. In Gambia, per capita exports in 1973 were the same as in 1965, but in 1974 their value nearly doubled. In Senegal and Togo, exports increased even more sharply between 1973 and 1974, but that was mainly a result of enlarged proceeds from exports of rock phosphate and phosphate fertilizers. Increases on the order of 50 percent were realized in Ghana, Sierra Leone, Somalia, Sudan, and Tanzania; and in Kenya, Madagascar, and Malawi the per capita value of exports approximately doubled between 1965 and 1973 and 1974. In Cameroon, the figure doubled between 1965 and 1973 and rose almost 30 percent between 1973 and 1974. Per capita exports also increased very substantially in Liberia and Zambia, but those increases were dominated by mineral exports.

## Domestic Sales

Rapid urban growth in recent decades has led to increased demand for purchased food, most of which has been satisfied through commercial sales by domestic producers.[19] Rough but plausible estimates made by FAO for the twenty-four countries included in their Africa Regional Study prepared in connection with the *Indicative World Plan for Agricultural Development* suggest that by 1962 between 40 and 45 percent of total agricultural output was marketed. And of the total marketed output, it was estimated that crop sales to domestic consumers accounted for slightly less than 50 percent and exports for slightly over 50 percent of the total (FAO 1968a:23). According to FAO's projections for 1985, based on the assumption of fairly high rates of urbanization and growth of nonfarm production and incomes, the per-

[19]According to an analysis published by FAO in 1968, imports provided only about 4 percent of total calories but 16 percent of total animal protein; these figures include some imports from other African countries. The same source indicates that imports represented only about 9 percent of the value of total food consumption but a quarter of the value of marketed output (FAO 1968a:35, 39). However, reliance on imports appears to have increased in recent years, although it is difficult to generalize because of the importance of year-to-year and intercountry variations in food imports. As noted in the earlier discussion of the food economies of tropical Africa, domestic production has apparently not kept pace with the growth of population, and a number of countries have responded to that situation by expanding their imports of cereals. This is especially true of the Sahelian countries, where the demand for imports was increased by sharp declines in domestic production and, because of the availability of food aid, foreign exchange constraints were not binding. The percentage increases between 1966 and 1975 were especially large in Mali and Mauritania—from 16,000 to 192,000 and from 2,000 to 122,000 tons respectively. (But in 1976 and 1977 Mali exported cereals.) A combination of unfavorable weather conditions and availability of food aid were probably also major factors underlying the sharp increase in Tanzania's cereal imports from 88,000 tons in 1966 to 461,000 in 1975.

centage of marketed output sold to domestic consumers would increase to 62 percent of total sales. This was related to a projected increase of 180 percent in total commercial production, for the domestic market and exports, between 1962 and 1985, by which time marketed output would account for nearly 60 percent of total farm output.

The approximations relating to commercialized production in 1962 are necessarily very rough because the statistics are incomplete and deficient. The projections for 1985 are even more uncertain. Moreover, there is great variation among countries in the extent to which agricultural output is marketed and in the relative importance of commercial production for domestic markets and for export.

Estimates of the imputed value of subsistence production as a percentage of total value of agricultural production have been published by Keith Abercrombie for eleven African countries, along with comparable figures for a few Asian countries and some high-income countries (Abercrombie 1965:2). For the period 1955–59, subsistence production in the African countries for which estimates were possible ranged from 28 percent of total agricultural output in Rhodesia and 41 percent in Uganda to 72 percent in Zambia and 75 percent in Guinea. In Rhodesia and Zambia, European farmers accounted for a sizable share of the commercialized production. For African farmers only, subsistence production in Rhodesia represented 76 percent of their total output and in Zambia, 91 percent (compared to the figures of 28 and 72 percent cited earlier for subsistence production as a percentage of total agricultural output). According to the 1970–71 census for Zambia, however, the percentage of production marketed had increased considerably. The census data for maize, the principal crop, indicates that 35 percent of the production of African farmers was marketed, and this was over two and one-half times the volume marketed by European farmers in that year. Abercrombie also gives estimates for 1960–63 for some of the eleven countries, but the change between the two periods was probably influenced more by changes in the relative prices of export commodities and domestic food prices than by changes in the relative share of output (in physical terms) that was commercialized. The relatively small share of subsistence production in total agricultural output in Uganda in the 1955–59 period was influenced considerably by the high prices then prevailing for coffee and cotton, the country's major commercial crops. In terms of current prices, the share of subsistence production in Uganda rose to 54 percent of the total in 1965, but if the calculation is made on the basis of 1960 prices, subsistence production still represented only 41 percent of agricultural output (Cleave 1974:24n).

In a comparative analysis of farm survey data for approximately fifty localities in ten countries, John Cleave found that the average percentage of farm output marketed by different survey groups ranged from less than 10 percent to more than 90 percent, and in half of the areas surveyed over 50 percent of farm output was marketed (Cleave 1974:Table 2.2, 14–30). Cleave notes that the areas selected for farm surveys may have been biased towards areas where commercial production is of more than average importance, and in any event such information from scattered surveys can only be suggestive.

There is considerable variation among major food crops in the proportion of output that is marketed. Estimates for Zaïre, for example, indicate that in 1959 only about 30 percent of the total production of maize and plantains was commercialized; and for manioc, the dominant staple food crop, only 17 percent of total output was marketed. On the other hand, rice was grown primarily as a commercial crop and nearly two-thirds of total output was marketed, and an even higher proportion of the production of truck crops was sold. But rice and truck crops accounted for only a small fraction of the value of farm output. The relatively small fraction of the major food crops marketed is rather striking, considering the extent of wage employment in the Congo at that time. In few countries in tropical Africa is reliance on wage employment as pervasive as it was in the Congo in 1960—nearly 1.1 million wage earners out of a total adult male population of 3.6 million (Drachoussoff 1965:166, 197).

Intercountry variations in the share of output marketed are, of course, related to large differences in the importance of export production as well as variations in domestic commercial sales reflecting differences in the size and per capita incomes of the nonfarm population. In addition, there may be substantial differences associated with variations in the importance of intrasectoral sales of agricultural commodities. Rough estimates by Hardie Park (1969), based on data from urban and rural consumer surveys, indicate that sales for domestic consumption accounted for nearly 60 percent of the farm value of marketed output in Nigeria in the early 1960s but less than 15 percent of the farm-level value of marketed output in Tanzania.[20] Some of the difference can be attributed to the fact that on a per capita basis, agricultural exports in Tanzania were nearly twice as large as in Nigeria—about $13 versus $8 in 1965. The extremely small volume of

[20]Park's estimates of the average value of purchased food as a percent of the value of all food consumed in all of rural Nigeria, based on data reported in Nigeria (1966), are as follows: starchy staple foods, 25.2%; meat, fish, eggs, milk, 90.5%; vegetables, fruits, nuts, 52.4%; oils and fats, 87.9%; salts and spices, 97.1%; kola, 88.6%; total food, 49.6%.

domestic sales in Tanzania was strongly influenced by the fact that some 95 percent of its population was still engaged in agriculture, as compared to perhaps 80 percent in Nigeria in the mid-1960s.[21]

## Agricultural Processing and Marketing Services

Processing of agricultural commodities by specialized firms is still of limited importance in the economies of tropical Africa, although it bulks large in their extremely small manufacturing sectors. For the substantial fraction of food consumption that comes from subsistence production, most of the simple processing is carried out in the farm household. The principal exception appears to be the widespread use of village mills for grinding maize and other commodities that are converted into meal or flour.[22] Although hand pounding of rice is fairly common, especially in the traditional rice-producing areas that extend along the coast of West Africa from southern Senegal to the middle of the Ivory Coast, a large fraction of the rice consumed is mechanically processed either by small village mills doing custom work or by larger establishments. In fact, the possibility of being relieved of the time-consuming task of pounding millets, sorghums, or other traditional cereals by hand is cited as one of the factors that has contributed to the rapid increase in rice consumption during the past quarter century. It is estimated that in Nigeria there were some 320 small rice mills and four large-capacity mills in operation in the mid-1960s (USDA/AID 1968:134). A large proportion of rice production in Nigeria is commercialized, and most of the crop appears to be mechanically milled.

It seems probable that the greater part of agricultural processing is still carried out by rural households using traditional hand methods. In addition to the processing of food for their own subsistence consumption, rural households perform much of the primary processing of export crops and of farm products destined for urban consumers. A high proportion of the labor time devoted to food production in tropical Africa is spent on postcultural operations such as shelling, cleaning, and grinding maize or processing

[21]This contrast between Nigeria and Tanzania seems to typify a general contrast between West Africa and eastern and central Africa. The greater importance of local markets and exchange in West Africa is of long standing. It is attributed mainly to the greater stimulus of long-distance trading contacts and a political structure capable of maintaining peace. It is also associated with the fact that rotational bush fallow is much more important than shifting cultivation in West Africa whereas the reverse is true in East and central Africa. That contrast, however, is probably more a consequence than a cause of the difference in the development of exchange and greater concentration of population (Morgan 1969:247–48, 254).

[22]Such mills are used to produce manioc flour in Zaïre. In West Africa, corn mills are often used in producing *gari*.

cassava roots into *gari* or cassava flour. Some of the most promising opportunities for increasing farm labor productivity lie in expanded use of a variety of fairly simple machines for carrying out these auxiliary operations.

On the basis of approximations that are necessarily very rough, FAO has estimated that the value added by agro-allied industries in 1962 was only $1.1 billion, compared to the estimated farm-level value of agricultural production (inclusive of primary processing), which was estimated at $7.7 billion. Such a low figure for value added by agro-allied industries is not implausible. Even urban consumers purchase mainly agricultural products that are little processed. The part of agricultural production that is exported accounts for a sizable but unknown fraction of the value added by agro-allied industries. The magnitude must, however, be fairly small. Most agricultural products are exported as raw materials that have undergone only primary processing at the farm or village level—fermentation and drying of cocoa beans; pulping, fermenting, washing, and drying of coffee beans; shelling of peanuts. And even the value added in crushing peanuts or palm kernels for oil is a small addition to the value of the primary product.

FAO's Africa Regional Report projects that by 1985 the value added by agro-allied industries will rise to $5.3 billion, compared to a projected increase in the farm-level value of production to $16.5 billion (FAO 1968a:20). Such projections can only be regarded as a very rough indication of the order of magnitude of possible changes, and the changes that have occurred since 1962 scarcely suggest that such a large increase in value added by agro-allied industries will be realized by 1985. There is no doubt, however, that with increased urbanization and rising per capita incomes, the value of agricultural products as purchased by consumers will embody substantially greater value added from processing. And even apart from changes in the degree of processing, the growing importance of specialized production for sale will mean large increases in the value of such marketing services as transporting, storing, handling, and packaging. Processing for export will also expand with increased processing of traditional exports and with the growth of exports such as canned fruit or juices.

Leaving aside primary processing of foodstuffs by farm households, processing for export is probably more important than for the domestic market. Estimates for Nigeria suggest that about 30,000 persons may be employed in export processing in firms employing ten or more persons—

well over a third of total employment in such firms.[23] Sawmilling and plywood firms, however, account for nearly half of that total. Among agricultural products, cotton ginning, palm oil extraction, and groundnut crushing bulk largest in terms of employment in Nigeria—and probably in Africa as a whole.

Palm oil is made from the perishable palm fruit in all of the oil-palm-producing areas.[24] Tea is another product that must be processed locally; factories for wilting and curing the freshly plucked leaves must be located in close proximity to growing sites, whether the tea is grown on estates or by smallholders.

For most of the other export products, the location of processing is optional and is determined by a combination of economic factors and governmental policies (especially the level of import duty imposed on the raw material as compared to imports that have undergone processing). The African producing countries have an important advantage in the lower cost of unskilled labor, and transportation costs will be lower if processing reduces the weight and bulk of the raw material. The saving in transport cost, however, is often considerably less than the reduction in volume because of special facilities required for handling the processed product. Moreover, the weight reduction may be slight when there is a market for the main by-products. European countries, for example, invariably import wheat because there is local demand for the bran as well as flour. The same applies to soybeans, with the sales of soybean meal being almost as impor-

[23]It has been estimated that in the late 1960s, employment in Nigeria in urban firms with ten employees or more totaled 76,000, that employment in smaller urban firms totaled another 100,000, and that about 900,000 persons were working in rural households engaged in manufacturing (Kilby 1969:17, 19, 138). An unknown but certainly sizable fraction of the latter two categories are also engaged in food processing, although many of the small enterprises would combine preparation of foods ready for consumption, e.g., couscous or *kenkey,* with vending the product. (Couscous, a ubiquitous product in the savanna areas of West Africa, are round balls made from finely ground millet or sorghum flour. *Kenkey,* dough balls made of cornmeal, is a popular and convenient food in Ghana.)

[24]Methods employed range from highly labor-intensive techniques carried out by farm households to the large-scale, capital-intensive mills used for plantation production. Peter Kilby (1967) provides an excellent analysis of the technical and economic characteristics of the four alternative processing techniques used in Nigeria. He reaches the important conclusion that the technically inferior labor-intensive practices remained dominant because of price policies of the Nigerian Marketing Board, which held down the producer price of palm oil to only about one-half the FOB export price. Because of the low price, the value of the incremental output resulting from the use of Pioneer mills or hydraulic presses that obtain a higher extraction rate was not sufficient to offset the lower cost per cwt. of the primitive technology, which used unpaid family labor and had much lower costs for depreciation and repair of equipment.

tant as the revenue from soybean oil. This consideration is less important when there is a domestic market for the lower-value by-product, but until very recently African oil processors have found little home demand for presscake from oil seeds, and export most of it to European countries where it is in demand as livestock feed.

The developed importing countries enjoy some offsetting cost advantages due to lower costs for capital, management, and skilled labor, and their processors usually realize economies of scale that are not possible for firms in Africa. Processors in an importing country may also have an advantage because of being able to shift from one raw material to another in response to changes in relative prices; and in some instances, notably in the production of cocoa butter and powder, the ability to produce to specific quality standards of individual customers may be important. In addition, differential tariffs that represent 100 percent or more of the gross value added in processing are commonly maintained by importing countries to protect their existing processing industries. However, the United Kingdom's policy of permitting duty-free imports of processed agricultural products as well as raw materials from Commonwealth countries has favored fairly rapid growth of a groundnut crushing industry in Nigeria since the early 1950s (Kilby 1969: 170–74). A sizable groundnut-crushing industry was established somewhat earlier in Senegal, made possible by duty-free imports of oil into France, although such shipments were subject to quotas until 1952 (Thompson and Adloff 1958:385).

Import substitution strategies have also given an impetus to the expansion of agricultural processing in a number of countries. Wheat consumption has been increasing more rapidly than consumption of any of the indigenous staple foods, and apart from Kenya, Ethiopia, and a few other highland areas, supplies are almost entirely imported. Until about 1960 imports were in the form of flour, with the exception of Senegal, Kenya, Rhodesia, and Mozambique, but since that time there has been considerable investment in flour mills in other African countries. In some instances the large capital investment required for a modern flour mill has been provided by foreign investors interested in access to a growing and protected market. Even though the foreign exchange saving is small and the amount of employment created very limited, investments in flour mills have been attractive to governments because they cater to an existing and sizable domestic market. Hence it is relatively easy to ensure profitable operation in spite of some technical problems, such as the lack of a local market for milling by-products.

Similar considerations have also prompted the establishment or expansion of domestic sugar industries, although these projects have naturally included the introduction of cane production as well as sugar milling. Technical problems encountered in the establishment of sugar industries appear to have been more serious than those encountered by the flour mills, especially in Ghana and Nigeria.

On the other hand, the sugar industries of Zaïre and Uganda are well established. Zaïre's production of 18,000 tons in 1955–57 supplied 68 percent of the requirements of the Congo customs union, which included Rwanda and Burundi, and Uganda's production of about 50,000 tons permitted exports to Kenya and Tanzania (Jones and Merat 1962). Between 1953 and 1964, Uganda sugar production increased by nearly 10 percent per year, compared to a 6.5 percent rate of increase in consumption, so that exports to the neighboring territories expanded considerably. Production in Kenya and Tanzania was at a very low level in 1954, but expansion of production since that time has been rapid. By 1972, production amounted to about 100,000 tons in both Tanzania and Kenya, compared to 132,000 tons in Uganda. But in 1974, Uganda's production was down to 62,000 tons whereas Tanzania and Kenya had reached 124,000 and 132,000 tons respectively. In Zambia, a parastatal sugar estate that first came into production in 1968 was supplying 70 percent of the rapidly increasing domestic consumption by 1971 (Lombard and Tweedie 1974:30).

Import substitution has been a consideration in the establishment of canning and bottling plants. The establishment of local manufacture of beer and soft drinks has been one of the most widespread and successful instances of industrial development in the years since World War II. Experience with the creation of factories for canning fruits, juice, and vegetables has been considerably less satisfactory. For a few products, for example tomato puree in Ghana, the domestic market, which has grown with the expansion if imports, is sufficient to justify full-capacity operation of at least one plant of economic size. In general, however, domestic demand for these products tends to be limited because of the low level of per capita income. Processing can obviously yield significant benefits to consumers in making perishable products available year-round and in providing an opportunity to buy products in a more convenient and perhaps more desirable form. But where fresh produce is available, the traditional practice of relying on it even though that means changing the menu with the season is likely to persist.

Owing to the limited size of internal markets and the diseconomies of very small canning factories, much of the expansion of production of canned foods is likely to be for export markets. Government encouragement of canning factories is also likely to be directed toward products that appear to have reasonably good export prospects because of the general concern with enlarging foreign-exchange earnings and the fairly high foreign-exchange requirements for imported machinery and for tinplate. Production for export, however, requires the ability to ensure regular supplies of the required quality at competitive prices, and this is not always easy. It also requires that the processing plants be able to obtain supplies of the raw produce on a regular basis. Experience in Ghana illustrates a number of these difficulties, notably that of obtaining a sufficient and regular throughput of raw materials to enable plants to operate at full capacity. Quality problems have been compounded by the difficulty of obtaining raw materials of uniform and suitable quality. Similar difficulties are reported for factories established in Nigeria for canning citrus fruit and meat. (See Reusse 1968 and Kilby 1969:185–86.)

There are a few products that merit special attention for nutritional as well as economic reasons. Plants producing "filled milk"—using imported dry skimmed milk and locally produced vegetable oil—either for sale as natural-type milk or for canning as reconstituted evaporated milk, seem to offer interesting possibilities (Reusse 1968:111–14). Milk products bulk large in the food imports of a number of countries, and locally produced evaporated milk could achieve a net foreign-exchange saving of about 40 percent. The health advantages of the expanded use of milk products in African countries are mixed. Although milk is of course an excellent source of high-quality protein, a shift from breast- to bottle-feeding will generally have adverse consequences on the health of infants and small children. Because of the high cost of commercial products, mothers often use mixtures that are too diluted to meet the nutritional needs of the infant. Even more important are the serious problems of diarrhea that arise with bottle-feeding when maintenance of high standards of sanitation is difficult. Continued breast-feeding also tends to increase spacing between children, but nutritionists advocate beginning supplementary feeding when infants reach six months. Products made from mixtures of locally available cereals and protein concentrates, particularly from the protein-rich presscake of groundnuts or cottonseed, would greatly facilitate the efforts to promote nutritional improvement during the "toddler" period of high risk when mother's milk is no longer sufficient. High-value uses for oilseed by-products would also raise the profitability of local processing, for reasons that were noted earlier. Improved availability in the savanna regions of products with a high content of vitamin A and other nutrients that are in short supply during the long dry season is another instance in which processing could make a notable contribution to nutrition. In West Africa, the palm oil available in the higher-rainfall coastal zones would seem to offer considerable promise because it is cheap and rich in vitamin A. Developing a market for such products will, of course, be difficult because of the meager purchasing power of rural households, especially in the savanna regions, and the problems of introducing new products into a traditional diet. But public support for research, pilot projects, and promotional activities to encourage this type of processing could yield substantial social returns to improved health as well as fostering the development of a manufacturing industry with good growth potential and based on domestic agricultural production.

## Farm-Supplied and Purchased Inputs

A fundamental characteristic of the agricultural economies of tropical Africa is the heavy reliance on "internal" inputs of land and labor and the extremely limited use of "external" inputs purchased from outside the agricultural sector.[25] It has been seen that the expansion of agricultural output during the past half-century was mainly the result of the introduction of higher-valued export crops superimposed upon a "horizontal" expansion of food crops to satisfy the needs of a growing population. In most areas there was large scope for bringing additional land into cultivation, and increased inputs of labor resulted from fuller utilization of the available "stock" of labor as well as from population growth.

Probably the most important change that has occurred in the use of purchased inputs has been the increased reliance on hired labor. In virtually all areas in which cash cropping has expanded there is considerable hiring of labor—either migrants from areas where income opportunities are restricted or members of local farm households who have a particularly limited supply of land and working capital. Although direct hiring of farm labor, usually for a daily wage, has become a common practice, indirect arrangements which do not involve cash payment are also widespread. Important examples are the various arrangements used by cocoa farmers in Ghana (Hill 1963: 187–92). *Abusa* and *abunu* laborers, who receive respectively one-third and one-half of the cocoa harvest, are still commonly used. Uchendu found at the time of his survey in 1968 that farmers preferred to hire laborers to carry out specific tasks in return for a daily wage,

---

[25]Land is supplied by the farm household in the sense that family labor, perhaps augmented by hired labor, clears forest or bush land to create fields that can be cultivated or planted to perennial crops.

but the practice was not widespread because many did not have the working capital to pay their laborers in cash (Uchendu 1969:22).

Use of inputs such as agricultural chemicals or farm equipment purchased from outside agriculture has increased considerably in recent years,
but from an exceedingly low base. Until lately, increased use of pesticides,
particularly the spraying of cotton and cocoa, has probably had the most
significant effect on agricultural production. Subsequently, however,
fertilizer use appears to have been increasing more rapidly than use of
pesticides. As recently as 1969, imports of fertilizer by the countries of
tropical Africa were only slightly greater than imports of pesticides. But
between 1969 and 1974, imports of manufactured fertilizer rose from some
$50 million to $180 million whereas pesticide imports rose from a little less
than $50 million to about $110. More rapid increases in fertilizer prices
probably account for much although not all of the difference in the growth
of imports, and fertilizer use has also been augmented by expanded domestic production (FAO 1976b). Although fertilizer consumption per hectare
of arable land and land under permanent crops rose substantially, from 1.8
kilograms of nutrients in 1965 to 4.9 kilograms in 1974, the level of use
was still extremely low (FAO 1976e). The most striking contrast is with the
developed countries, where the consumption of fertilizer per hectare already exceeded 70 kilograms in 1964 and surpassed 100 kilograms in
1972. A more significant comparison, however, is with the developing
countries of Asia, where the spread of high-yielding varieties of rice and
wheat stimulated rapid expansion of fertilizer use from 6.4 kilograms per
hectare in 1964 to 21 kilograms in 1974. African countries that have
experienced a substantial increase in fertilizer use are those in which high-
yielding varieties of maize have been introduced. The most striking examples are Kenya, where per hectare consumption rose from 11.5 to 30.0
kilograms between 1964 and 1974, and Zambia, where consumption rose
from 2.2 to 10.7 kilograms. The lower figure in Zambia reflects the greater
concentration of fertilizer use in commercial farming areas such as
Mazabuka District (see Chapter 5). Fertilizer use in Rhodesia—33 kilograms per hectare in 1964 and 62 kilograms in 1974—is by far the highest in
tropical Africa; but this is influenced by the importance of tobacco as well
as the extent of commercial production by European farmers.

Purchases of other inputs, including hand tools, ox-drawn equipment,
tractors, fuel and lubricants, seeds or other planting material, and livestock
feed, amounted to about $110 million in 1962 for twenty-four countries
covered by the FAO Regional Survey (FAO 1968b:545). It was estimated

that purchases of fertilizers and pesticides in that year amounted to about $30 million, so total purchases of inputs from other sectors were about $140 million, less than 2 percent of the estimated value of gross farm output in the twenty-four countries. Although outlays for purchased inputs have increased considerably since 1962, it seems unlikely that they amount to much more than 5 percent of the value of gross farm output in the countries of tropical Africa. This is in sharp contrast with developed countries, where high levels of agricultural productivity and output depend upon the use of large quantities of yield-increasing and labor-saving inputs obtained from the industrial sector. In the United States, for example, outlays for current inputs plus depreciation and other consumption of farm capital exceeded 60 percent of gross farm income as early as 1960 (USDA 1963:45).

The use of resources for irrigation purposes has been limited in tropical Africa. Considering the region as a whole, there is an overwhelming predominance of rain-fed agriculture. The only notable exception is Madagascar, which reflects the long and powerful influence of its Asian cultural tradition, although there are a few continental areas of limited extent where irrigation has been practiced for many years. More recently there has been some development of minor schemes for irrigation and of a few large-scale projects. It has been estimated that for twenty-four countries south of the Sahara, only 1.8 percent of the total harvested area is irrigated; and if Madagascar, where just over half the harvested area is under irrigation, is excluded, the figure falls to .5 percent (FAO 1968a:130).

## Agricultural Expansion and Overall Economic Development

In view of the predominance of agriculture in the economies of tropical Africa, changes within the agricultural sector are bound to have a major influence on overall economic growth. During the past half-century the expansion of agricultural exports has clearly been the most powerful force leading to economic change, except in a few countries where mineral exports have been dominant. Enlarged production of the various export crops has both stimulated and helped to finance expansion of infrastructure, including an improved transport network, ports and harbors, the supply of electricity, and communication facilities. Earnings from export crops have also been important in financing a tremendous expansion of educational facilities and more modest but significant progress in establishing other governmental services such as agricultural research stations.

Experience in Africa, as in other developing countries, has demonstrated, however, that although the creation of capital and institutional infrastructure is a necessary condition for economic growth and development, it is by no means a sufficient condition. And there is today widespread concern because the rather impressive growth of export production has not given a stronger impetus to the development of local manufacturing industries. The reasons why the growth-promoting interactions between agriculture and industry have been disappointing are complex, and can only be touched upon in this study.[26]

The factors that have restricted the importance of forward linkages have already been examined in assessing the growth of agricultural processing industries. It is obvious that the backward linkages associated with the expansion of agricultural output have been weak, given the limited use of purchased inputs.

Insofar as increased cash incomes have permitted expanded use of purchased inputs, this demand has been directed mainly toward hiring labor or toward imports of farm equipment, fertilizers, or pesticides. Frequently, the choice of policies for both agriculture and industry has been poorly suited to maximizing the positive interactions between agricultural and industrial development and to fostering rapid growth of output and employment opportunities in the two sectors. These policies have adversely affected the growth of effective demand for inputs and consumer goods by farmers and have favored the growth of inappropriately capital-intensive industrial firms. In particular, little attention has been given to the interrelated objectives of: (1) increasing agricultural productivity through the introduction of improved but low-cost farm equipment suited to the needs of small-scale farmers, and (2) promoting the establishment and growth of rural-based workshops manufacturing such equipment by technologies which rely heavily on labor and other indigenous resources. As stressed in later chapters, increases in farm productivity also require major efforts to generate and diffuse biological and chemical innovations such as high-yielding and fertilizer-responsive varieties of maize and other staple food crops. However, expanded use of hand-operated or animal-powered equipment and other implements that are well designed but cheap and simple enough for widespread use and local manufacture is of special significance. Equipment innovations with those characteristics can stimu-

[26]But see Scott Pearson and John Cownie (1974) for an examination of the relationships between export expansion and development in a number of African countries. For a general analysis of the interactions between agricultural and industrial development, see Johnston and Kilby (1975).

late the development and spread of metalworking skills and capacity as well as contributing to higher productivity in agriculture.[27]

To be sure, the presence of tsetse and trypansomiasis has precluded the introduction of animal-drawn equipment in a number of areas, and cultivation of the tree crops which bulk large in Africa's exports is difficult to mechanize. Probably more important, however, has been the widely held view that African farmers could and should make the transition directly from the hoe to the tractor. The enormous rise in the cost of fuel and substantial increases in the price of tractors and spare parts has, however, encouraged a reexamination of that view; and it is now more widely recognized that broadly based increases in farm productivity will require a greater emphasis on inexpensive types of equipment that smallholders with limited cash income can afford. Concern with the problems involved in expanding jobs and improved income-earning opportunities for a rapidly growing population of working age has also directed attention to the need for equipment innovations which ease labor and energy bottlenecks and improve the timeliness of farming operations without curtailing the expansion of opportunities for productive employment in agriculture. But research and development activities to identify appropriate and promising equipment innovations have generally been neglected, and even less attention has been given to measures to encourage wide adoption of such equipment and to foster local manufacture of items which have good demand prospects.[28]

Although many countries have emphasized industrial expansion, the import substitution strategies pursued have often led to adverse as well as favorable consequences. High levels of protection have had the effect of turning the internal terms of trade against agriculture, so that the costs involved in maintaining high prices for domestic manufactures have been borne largely by the agricultural sector in the form of lower export prices (because of overvalued exchange rates) and of the higher prices that farmers have had to pay for imports and domestically produced consumer goods or inputs. Heavy taxation of exports, including the de facto taxation levied by marketing boards, which have often paid farmers much less than would

[27]Nathan Rosenberg (1976:chap. 5) and Peter Kilby (Johnston and Kilby 1975:chap. 8) have emphasized that the strengthening of technological capacities in metalworking and the development of an indigenous capital-goods-producing sector have especially important feedback effects, e.g., in enabling a developing country to adapt and create technologies that are appropriate to its factor endowment.

[28]For an examination of these issues in the Kenya context, see Sidney Westley and B. F. Johnston (1975), which also includes an account of the activities of Tanzania's Agricultural Machinery and Testing Unit.

have been warranted by world prices, has had similar consequences. Moreover, a variety of factors, including government wage policies, minimum wage legislation, and trade union pressures, have often led to large wage increases for some urban workers. Thus income differentials between those employed in the modern sector, as compared to rural incomes and earnings of workers in small-scale firms in the traditional manufacturing and service sectors, have increased greatly. Other economic policies, most notably the fixing of interest rates at artificially low levels for those fortunate enough to be able to receive loans from government agencies or other institutional lenders, have commonly resulted in the underpricing of capital relative to labor.

As a result of these interrelated factors, investment has all too often been concentrated in capital-intensive activities and the growth of nonfarm employment typically has been slow. (See, for example, Todaro 1971 and ILO 1972:chap. 6.) More generally, the process of transforming the economic structure of the predominantly agrarian economies of tropical Africa has not only been slow, but in addition has frequently not had a pervasive effect in stimulating the increased specialization and widespread improvement in technical and entrepreneurial skills required for broadly based economic growth. The limited development of rural-based industries has already been stressed. The slow expansion of nonfarm employment and the restricted growth of rural purchasing power have also dampened the rise in domestic commercial sales and limited increases in intrasectoral purchases of food. Those adverse effects on the commercialization of agriculture have also impeded the increases in productivity that are stimulated by the growth of specialization in agriculture.[29] Some of the other issues pertinent to strengthening domestic food-marketing systems are examined in the next chapter.

Many factors have been responsible for the restricted growth of rural purchasing power, and their relative importance has varied over time and from country to country.[30] A major conclusion that emerged from the earlier analysis of export expansion was that the twofold increase in the volume of agricultural exports between 1950 and 1969 led to an almost

[29]In his study of the food-crop economy of Nigeria, Godwin Okurume places considerable emphasis on the productivity increases in food production that can be expected to result from increased division of labor and specialized production in areas that have a comparative advantage for food-crop production. He also notes that specialization in export-crop production involving increasing dependence on food purchases will be more likely if an efficient food market exists (Okurume 1969:21, 28–30).

[30]For an analysis of why real farm incomes in Uganda failed to improve between the 1930s and 1960s despite impressive expansion of agricultural exports, see Vali Jamal (1976).

equally large increase in export proceeds. But even though the overall decline in prices for tropical Africa's major agricultural export products was small, there was a considerable deterioration in terms of trade because of the rise in import prices during the twenty-year period. Moreover, the increase in per capita value of agricultural exports was considerably less than the twofold increase in volume. The total population in the countries of tropical Africa increased by something over 50 percent during that period, and in most countries the growth of the farm population was also very large. For the period 1960 to 1970 it is estimated that a 29 percent increase in Africa's total population was associated with a 20 percent increase in the agricultural population (FAO 1976a:35).

It was further noted that the expansion in the volume of exports leveled off markedly beginning in 1965, and with the exception of tea this relative stagnation in agricultural exports continued through 1975. For a number of crops a sharp increase in world prices in 1973 and 1974 led to a substantial rise in the value of exports. But with the exception of Nigeria and Gabon—which have benefited from the fivefold increase in petroleum export prices since 1973, when the Organization of Petroleum Exporting Countries (OPEC) began to force up prices by curbing production and exports—the terms of trade and the balance-of-payments situation of all of the countries of tropical Africa have worsened. High rates of inflation in the industrialized countries have also led to large increases in the import prices of most manufactured products, which have been an important factor in the deterioration in the terms of trade.

The effects of rising import prices on the terms of trade of many countries in Africa and other developing countries have stimulated a great deal of discussion of measures aimed ''at achieving stable conditions in commodity trade at levels remunerative and just to producers'' in developing countries (FAO 1976c:13). The May 1976 session of the United Nations Conference on Trade and Development, for example, adopted a statement setting forth the objectives, pricing and stockpiling policies, and timetable for an Integrated Programme for Commodities. For many years there has been considerable interest in international commodity agreements aimed at avoiding excessive price fluctuations and preventing deterioration in the terms of trade of primary exporters. During the 1960s, the International Coffee Agreement appears to have had some success in forestalling any sharp decline in coffee prices, but this was a mixed blessing for some African producing countries, which were obliged to limit the rate of expansion of coffee exports in accordance with quotas negotiated under the Agreement. In 1972/73 coffee export quotas were abandoned, and the extension of the agreement did not provide for quotas or other economic provisions. New international agreements for both coffee and cocoa were negotiated in the fall of 1975, but it is too early to assess their effects on coffee and cocoa prices and exports.

It seems reasonably clear that stagnant or declining prices for a number of commodities have contributed to the slowdown in expansion of exports since 1965, just as the sharp rise in export prices in the early 1950s was a stimulus to rapid expansion. It would be wrong, however, to attribute the recent slowing of the rate of expansion of agricultural exports entirely to the unfavorable price trends for certain commodities or to the more general deterioration in the terms of trade experienced by Africa's agricultural exporters. There has been considerable variation among countries in the quantitative changes in their agricultural exports as well as in the impact of price changes, and in some instances supply conditions appear to have been primarily responsible for the sluggish growth or the decline of certain export industries. As noted earlier, the export performance of some countries appears to have been affected adversely by unfortunate trade and pricing policies as well as by domestic disturbances.

An important example is the decline in Nigeria's palm oil exports. One significant consequence of the underpricing of palm oil resulting from the large de facto tax collected by the government marketing board is that the Nigerian economy did not reap the potential benefits of new high-yielding varieties of oil palm. It also delayed adoption of improved processing techniques that would increase the production of palm oil because of higher extraction rates. More determined and more appropriate policies to stimulate increased production could have led to an expansion of exports based on significant increases in productivity; hence Nigeria's single-factoral terms of trade would very likely have been improved even though the barter terms of trade declined.[31]

The tremendous expansion of Nigerian petroleum exports in recent years has improved Nigeria's balance of payments position so drastically that expansion of foreign exchange proceeds from agricultural exports is no longer a priority concern. Even so, Nigeria still confronts the difficult problem of absorbing a rapidly growing labor force into productive employment and ensuring broad participation of the population in the process of economic growth. Given the large fraction of its population which still

---

[31]The barter terms of trade simply measure the change in the ratio between an index of export prices and an import price index. The single-factoral terms of trade correct the barter terms of trade for changes in productivity in producing exports.

depends on agriculture for work and income, problems of agricultural development are still of central importance.

Other countries, most notably Malaysia but also the Ivory Coast, have vigorously exploited the improved possibilities for low-cost production of palm oil. The rapid increase in palm oil shipments entering world markets in the early 1970s led to a considerable widening of the price spread between palm oil and other vegetable oils. Thus the price differential between palm oil and soybean oil, which averaged only $11 per ton during the 1960s, increased sharply in 1970 and 1971; in the latter year, the CIF price of palm oil in Europe was $47 per ton less than soybean oil. Although there is in general a high degree of substitutability among vegetable oils, the short-run price elasticity for an individual oil is apparently fairly low when importers are absorbing a sharp increase in exports—one of almost 30 percent in palm oil exports between 1970 and 1971 (Lynam 1972:8–10). With strong import demand and a sharp price rise for fats and oils in 1973, the differential between palm oil and other vegetable oils narrowed sharply, though it widened again in 1974 and 1975 when exports of palm oil again increased rapidly. In the longer run, however, United States producers of soybeans, West African groundnut producers, and other exporters of fats and oils will presumably experience price changes in response to variations in world supply and demand relationships for vegetable oils that will not differ greatly from the price changes affecting palm oil.

The relatively low level of prices of vegetable oils in the 1960s and early 1970s have, however, had an especially severe impact on some of the fragile African economies that depend very heavily on groundnuts as a major source of cash income and foreign exchange earnings. A monograph by Yves Pehaut on groundnuts in Niger notes that the positive stimulus to economic growth resulting from substantial expansion of groundnut production and exports in the 1960s had been limited because of declines in export prices. Niger and other francophone countries were able to export to France at prices considerably above world levels until 1968 when the "Common Agricultural Policy" of the European Common Market came into operation, so that the reduction in producer prices has reflected this policy change as well as the general downward trend of vegetable oil prices. Pehaut thus speaks of a "brutal" deterioration in Niger's terms of trade; the price of groundnuts relative to cement in 1968 was less than one-half its 1962 purchasing power (Pehaut 1970:101). As previously noted, the future outlook for peanut oil prices is not bright because of increasing competition from low-cost production of palm oil. Thus the general economic problems of the Sahelian countries related to their limited resource endowment and the high transport costs which characterize the landlocked countries are likely to be exacerbated by unfavorable terms of trade for their major export crops. Because of their small population and their small "economic size," in spite of their relatively large land area, the countries of the Sahel also epitomize a more general problem: the difficulty of achieving satisfactory industrial expansion and structural change in economies so small as to limit the scope for specialization and for realizing the economies of scale which are often important in manufacturing.

Given the benefit of hindsight, it is easy to criticize some of the economic policies adopted by African countries during the past decade and a half. It needs to be stressed, however, that the developing countries of tropical Africa face unprecedented problems as well as unprecedented opportunities. Because of the large body of science-based technologies that have been evolved in economically advanced countries, they can economize on the time and investment of resources required to produce the scientific and technical knowledge needed to raise economic productivity and output. This opens up the possibility of very rapid growth; but the large number of options that are potentially available make choice difficult. And there is a strong possibility that borrowed technologies will be ill suited to conditions in economies characterized by severe shortages of capital and foreign exchange and limited technical, entrepreneurial, and organizational skills and knowledge. Furthermore, the explosive growth of population and labor force during the past two or three decades compounds the difficulty of transforming the economic structure of these predominantly agricultural economies.

# 3. The Internal Market and Commercialization of Farming

African agriculture had experienced only modest increases in productivity and those mostly caused by shifts from less to more highly valued crops during a time when aggregate agricultural output was increasing rapidly. Total production grew because of the introduction of new crops and because resources of land and labor were transferred from their former employment (or unemployment) into farming. The expansion of output also depended on technical and organizational changes in the marketing and evacuation of the products and their transportation to the markets of Europe and North America.

It might have been expected that the specialization in production implied by growth of the new export crop industries would spread throughout agriculture to permit a general lowering of costs (increase of productivity). It has done so but only to a very limited extent, primarily because of the self-sufficiency of rural African communities. Much of the population is still occupied with producing food for its own consumption and relies very little on the market for its basic food needs. Production for export accounts for a substantial part of total farm cash income and may in some countries make up a large part of the total value of farm production.[1] Opportunities for further increases in production for export still exist, but the greatest potential expansion of market demand will be at home, as more and more African families obtain their principal food requirements in the market. The internal market, then, becomes of major importance if production for sale is to continue to grow.[2]

Specialization facilitates the substitution of more productive inputs

[1]Estimates of the proportion of agricultural production exported that can be derived from Tables 2.1 and 2.5 tend to be high. The weak statistical base causes underestimation of production of the food crops that are consumed domestically, and differing methods of valuation—farm prices for production, FOB prices for exports—undoubtedly result in overestimates for the proportion of farm output that is exported.

[2]Uma Lele points out that official pricing policies have favored export crop production at the expense of food crop production. Because of such policies of the Lilongwe Land Development Programme in Malawi, "gross return from an acre of tobacco is three times that

supplied by modern biological, chemical, and mechanical industries—the products of the new science-based technologies—for less productive traditional inputs of farm origin. Higher-yielding seeds developed by modern breeding techniques replace those selected by the farmer from his own crop or purchased from his neighbor. Cheaper, more effective, and more convenient nitrogenous fertilizers supplement or replace barnyard manure, night soil, and compost. Pesticides from the chemical industries are used to suppress and destroy insect pests that are all too often immune to traditional methods of control, and easily applied herbicides make unnecessary the tiring tasks of hoeing and pulling weeds. Finally, machines of many kinds, powered by draft animals or internal-combustion engines, reduce the physical burden and shorten the time of plowing, planting, cultivating, and harvesting, and of drying, threshing, shelling, husking, peeling, grinding, and pounding to prepare the harvest for use as food or industrial raw material. In these stationary processing activities, power may come from steam, electricity, or water as well as from draft animals and internal-combustion engines. Some reduction in time and effort may also be achieved by hand- or pedal-operated machines and by improved hand-powered farm implements.

It is not only by sharing productive activities with town-based industries, however, that specialization fosters increased productivity. Lower costs and greater economic output also result from the more efficient organization of production that can often be achieved without altering the technical coefficients of physical inputs. By engaging in a narrower variety of activities, the farm worker becomes more efficient in performing them and the farm decision-maker learns how to combine the inputs he controls more effectively.[3] The specialization in farm production that is made possible by exchange of goods and services also permits fuller exploitation of natural advantages in production. It can lead to man-made advantages, too, deriving from the provision of supporting and complementing facilities and activities like warehouses, railroads, highways and trucks, ports, posts, telecommunication, marketing specialists, organized market meetings, and credit institutions. Each of these requires a certain minimum volume of

---

from an acre of maize, despite the fact that new technologies exist for maize'' (Lele 1976: 265). Cyril Ehrlich reminds us that British civil servants in Africa envisaged economic growth "essentially in terms of the promotion of a narrow range of export staples, while imports mainly took the form of consumer goods which were taxed for revenue, rather than for the promotion of infant industries'' (Ehrlich 1973:649).

[3]Mechanical and managerial skills can of course be learned from others and are appropriate subjects of agricultural extension education. But they can also be self-taught, and are more likely to be so as the producing unit becomes more specialized.

produce and services offered or demanded if it is to operate at reasonable cost. The existence of such marketing, storing, and transporting facilities can often create for a particular locality a far greater competitive advantage than would result from natural causes. Production that is specialized in location can purchase the inputs it needs more cheaply and sell its output at higher prices than production that is dispersed.[4] The implications of this for agricultural development strategies can be politically painful. Areas of specialized production will inevitably be more prosperous than those of widely varied outputs, which is simply another way of saying that participation in a market economy can bring greater economic rewards than are possible in a truly subsistence economy.

But the division of labor is limited by the extent of the market. To the extent that farmers face a constraint on the amount of food and fiber that they can sell without a disastrous decline in price, they are discouraged from purchasing inputs such as farm chemicals, farm tools, and transport equipment that will increase their output, unless the resources so released can find rewarding alternative employment. Increased purchases of farm inputs without an equal or greater increase in farm receipts implies a decline in the amount of money that farm households can allocate to consumer goods, school fees, and savings.

To say that farmers do not buy fertilizer because it will not yield a net return, or that they will not apply it to a crop that is not grown for sale, is not the same as saying that they do not have the money to do so. The farm household may very well have cash resources that could be diverted to the purchase of productive inputs, and yet not so employ them because the reward is too small.[5] Or the total cash income may in fact be so small that it barely pays the taxes and permits only those purchases that are regarded as essential. If the constraint on purchase of production inputs is lack of markets rather than lack of cash, a program to advance production credit to farmers cannot be expected to have much effect on output. In either event, the marketability of the crops is of critical importance.

The extent to which farmers reject productive innovations offered by the extension officer because they do not have the money is not clear. Small outlays for purchased inputs can often be explained in large measure by the limited availability of profitable innovations. C. J. Doyle states that "small-

---

[4]This, of course, is one reason why plantations are established.

[5]Walter Elkan says (1976:694): "These resources are *not* confined to the savings out of farm incomes, but can be—and often are—supplemented by the profits of trade, the remittances from town-dwellers, and dry-season activities of members of the family who have turned themselves into village blacksmiths or carpenters."

holder credit surveys provide no clear evidence that shortages of institutional credit critically affect the decision to adopt," and cites studies by C. K. Brown (in Ghana), Josef Vastohoff (in Kenya), and John de Wilde's general study in support: "Certainly, for many simple innovations the African smallholder appears to be able to raise the small sums of cash required, either by drawing on his own capital or borrowing from relatives" (Doyle 1974:69). Guy Hunter, too, says there is "some detailed evidence that, in many situations, a feasible, profitable, and marketable innovation will be grasped by farmers without the need for subsidized official credit" (Hunter 1973:240).[6] And A. F. Bottrall, on the basis of a review of numerous recent studies of farm credit, including the twenty-volume *Spring Review of Farm Credit* published by the United States Agency for International Development in 1973, says, "The lessons from this evidence are that credit is useless as a stimulus to innovations when farmers are unwilling to adopt a new technology or have poor incentives to do so; and that when the necessary motivation has been provided . . . credit is usually unnecessary at the adoption stage" (Bottrall 1976:359). Expenditure accounts reproduced in this chapter from various studies seem to confirm Doyle's position.

On the other hand, C. Okali and R. A. Kotey found in a resurvey of Akokoaso, Ghana, in 1970 that credit was much less readily available to cocoa farmers than it had been at the time of W. H. Beckett's study thirty-five years earlier, while at the same time spraying for capsids had increased the need for credit. (Okali and Kotey 1971; Beckett 1944). Farmers, too, vary considerably in the resources they can command, and the well-to-do farmer may find innovation much easier than less affluent farmers do, as P. S. Zuckerman (1977) found in western Nigeria. There are obviously significant interactions among a country's economic structure, average cash farm receipts, the adequacy of the farm research base, and the nature and availability of innovations; they become increasingly important with the growth of specialization and of interdependence between agriculture and the other sectors.

## Precolonial Internal Trade

All societies are characterized by shortages; isolated societies produce surpluses only unpredictably and unintentionally. Economic exchange

---

[6]Hunter also points out that "to this day, despite lower rates of interest in official schemes, many farmers find merchant, moneylender, or relative a more convenient lender" (Hunter 1973:240).

probably first arose when members of one community suddenly found themselves possessed of more of some commodity than they could use, but that they could swap with members of another community who were eager to share in the first community's surplus. Or it could have arisen when a community found itself so desperately short of food or water that it was willing to swap a less desired good for it.[7] It was only later that farmers began to produce crops in excess of their own needs in order to exchange them for goods of which they did not have enough, thus improving their own well-being and that of their trade partners too. It must not be forgotten that this is the function of economic exchange—to improve the satisfactions of all who participate in it by reallocating products in accordance with consumers' preferences.

It is not known when economic exchange began in tropical Africa, but there is strong reason to believe that long-distance trade dating back 2,000 years or more was a principal means by which new productive methods spread throughout the continent, whether it was knowledge of ironworking emanating from the north, the banana-plantain spreading from the east, or rice and yam culture fanning out from West Africa. By the time Europeans first penetrated the interior of tropical Africa, an ancient caravan trade linked the western Sudan with the Mediterranean world and traffic in gold, ivory, and slaves flowed to trading stations on the shores of the Atlantic and Indian oceans.

Caravan routes of western Africa are best known and may have been most highly developed because of the relative proximity of the Sudanic kingdoms to the markets around the Mediterranean. E. W. Bovill's *Golden Trade of the Moors* (1958) treats primarily of this north-south trade across the Sahara. But well-established routes also ran east and west through the Sudan from the Atlantic Ocean to the Nile, and probed south through the forest to the upper Guinea Coast. Elsewhere in Africa there was an ancient trade from the mines of Katanga and Monomotopa, active salt trade throughout various parts of East Africa and the Congo Basin, and probably a very early trade in iron manufactures.[8]

Commodities moving in long-distance trade must possess high value

[7]Margaret Hay tells how the Luo on the northern shores of the Kavirondo Gulf in present-day Kenya were driven to expand their external trade during the 1890s when rinderpest swept through their herds: "This drive to rebuild livestock holdings through exchange was partly responsible for the expansion of external trade during this period" (Hay 1972:105).

[8]The series of collections of carbon 14 dates for Iron Age sites in sub-Saharan Africa that have been published in the *Journal of African History* show a surprising simultaneity in the earliest appearance down the eastern side of the continent, from Uganda to South Africa, at about 400 A.D., suggesting a rapid transmission of ironworking technology. This may have

relative to their portability, particularly when transport is principally by
human bearers. In an earlier period the object of such trade was mostly
minerals or products of the hunt—gold, copper, iron, and salt, or ivory,
ostrich feathers, wild animal skins, fish, and captive humans—but some
products of agriculture, like kola nuts, and manufactures, notably of iron
and leather (perhaps pottery as well), were also transported over long
distances. The trade in salt, fish, kola, pottery, and iron manufactures was
confined almost entirely to the continent itself. Gold, ivory, feathers,
skins, slaves, and leather goods were shipped, in early times, to the
Mediterranean, Arabia, and Asia; and copper may have been. It was unusual, however, for staple foodstuffs to travel very far. This is not to say
that the staples were not marketed at all; the great caravans crossing north
and south across the Sahara and east and west through the Sudan required
provisions, and markets to supply them grew up in places as different as
Port Loko in Sierra Leone, Timbuktu in Mali, and Kano in Nigeria.[9]
Commerce followed similar trade routes elsewhere in Africa, but the volume of trade was probably smaller. It is known, for example, that caravans
of slaves carried copper from the mines of Katanga and Zambia to ports on
the Indian Ocean. Gold was brought out from the Zambezi Basin at an
early date. There was also a regular commerce in ivory to Zanzibar from
the shores of Lake Tanganyika. These caravans, too, had to be fed, and
Andrew Roberts reports how merchants traveling through Gogoland carried with them iron hoes of local manufacture with which to buy food, and
slaves to buy ivory; salt from Katwe on Lake Edward or the Uvinza mines
between Tabora and Ujiji could also be used to buy food. But in the 1880s,
"One Arab at Tabora, Salim bin Sef, had abandoned trade for agriculture:
he employed a great many slaves in cultivating extensive plantations of
cassava and grain, but seems to have had no African counterparts"
(Roberts 1970:52–53, 65, 69). With the establishment of European rule at
the beginning of the nineteenth century, African farmers found new markets for foodstuffs in the requirements of administrative and military staffs,
merchants, and railroad builders.

There was also a local trade based on regional specialization in production. Jack Goody believes that mercantile activity in most of tropical Africa

resulted from migration or conquest, but commerce is certainly a possible explanation (B. K.
Jones 1971; Phillipson 1975). In the nineteenth century there was a well-documented trade in
iron goods over parts of this area (Gray and Birmingham 1970).

[9]C. W. Newbury (1971; 104) says that "coast trade was not a substitute for Sahara-
savannah trade"; and Claude Meillassoux (1971: 51) thinks that it probably never exceeded
the interior trade in value.

"was not vastly different from that of medieval Europe" (Goody 1971:24). In West Africa, local manufactures were an important stimulus to internal trade and closely resembled those of preindustrial Europe—cloth and clothing, metalware, ceramics, building materials, and processed food. The manufacture of cotton cloth was especially important (Hopkins 1973:48). Local manufactures along with agricultural products made up the major part of internal trade in West Africa and "assured the complementarity of inter-regional exchanges." Laboring slaves obtained through trade contributed importantly to the production of salt, gum, grain, cotton and woven materials besides serving as "farm labourers, cattle herders, well-diggers, miners, weavers, porters, boatmen, and ferrymen" (Meillassoux 1971:53). Long before they were conquered by the French in 1913, the Batéké, who lived north of Brazzaville, supported well-developed fixed markets to which came fishermen like the Bafourou and the Likouba, loading their great dugout canoes of two to five tons burden with manioc to be traded down the Alima River and up the Congo River for distances of up to fifty miles. It seems quite likely that fishermen generally were among the first merchants in riverine areas where their command of transport gave them a great advantage over the sedentary farmers.[10] G. Bruel, speaking about the possibility of developing production for export from French Equatorial Africa in a conference on French colonial agriculture held in 1918, said that all of the tribes of fishermen and traders along the rivers, like the Boubangui, the Banziri, and the Sango farmed very little and bought manioc, maize, and millet from the land people, leading to the creation of important markets a kilometer or two from the banks of the river outside the zone of annual flooding (Bruel 1920:129–30).

In West Africa, local trade appears to have been more vigorous and better organized than elsewhere in Africa, with markets that met regularly and were well attended, although similar markets were reported by early European and Arab traders in various parts of the Congo Basin and in the areas drained by the Zambezi River. Elsewhere, exchange was more episodic, as when the "Gikuyu, after collecting their trading goods, would send for their . . . friends in Masailand . . . to meet the traders at the frontier and conduct them into the country" (Kenyatta 1962:66), or the Masai in Tanzania "traded with their separate neighbors largely, or almost entirely perhaps, through the women of either side exchanging pastoral produce . . . for agricultural and forest produce" (Gulliver 1962:432). F. E. Bernard

[10]Just as the Wagenia and the Lokele are the principal merchants in the markets of Kisangani.

says that in nineteenth-century Meru, Kenya, "almost any item or good that any individual desired for everyday living could be self-fabricated or produced," but specialized production of honey, mead, salt, tobacco, a masticatory plant called *miraa,* and the products of blacksmiths, potters, and weavers of basketry were objects of trade among communities. "There is no record of exchange of other agricultural commodities" (Bernard 1972:42–43).

In many parts of tropical Africa, particularly the Congo Basin and areas bordering it on the north, east, and south, the arrival of the white man—whether European, Egyptian, or Arab—or of his black servants led to the dissolution of the marketplace, impressment of women traders, and a general disruption and breakdown of internal trade both in manufactures and in foodstuffs.[11] As the European states managed to achieve political control of African lands, they frequently granted monopolies for the collecting and marketing of lucrative export crops, while at the same time forbidding indigenous trade. The excesses of rubber concessionaires in French and Belgian territories are the most notorious aspect of these trading monopolies; according to Bruel, some of the concessionaires in the French Congo even went so far as to ask that manioc be considered as a product of the soil, thus making it subject to their control, and the Batéké forbidden to sell it to private traders. (The Bafourou paid the Batéké a little over a franc for a basket of manioc that they could later resell for 5 francs; Bruel 1920:130). This request was denied, but there must have been a great many instances of local merchants being driven from their traditional pursuits. Private trading concessions persisted in central Africa until the 1960s, and public monopoly of trade in export crops, sometimes of domestically consumed crops, is still a dominant feature of most of tropical Africa.

Throughout much of the colonial period there tended also to be a general lack of sympathy with commerce, an "upper middle class snobbery toward trade and manufactures" Charles Wilson has called it, that impeded the development of an indigenous entrepreneurial class and put barriers in the way of the free movement of goods and prices. This obstructionism from government, particularly in eastern and central Africa, makes all the more remarkable the achievement of African merchants after World War II (cf. Wilson 1965; Ehrlich 1973).

[11] Samir Amin provides an account of how in Senegal in the early years of this century the African merchant class, which had earlier been strongly supported by the French authorities, was undermined in favor of colonial businesses and the Lebanese and poor whites who were their agents (Amin 1971:361–76).

## Rural Productive Activities

We are accustomed to think of national economies as made up of populations that live in rural areas and earn their living by farming and of other populations in towns and cities which are employed in nonagricultural activities, principally manufacturing and the provision of services. This familiar view that identifies the agricultural sector with rural residence and the nonagricultural sectors with town residence is something of an oversimplification when applied to Western countries. It can be very misleading when applied to the tropical African countries.[12]

Although it is true, as stated earlier (see Table 2.1), that approximately 80 percent of the economically active population farm, this should not be taken to mean that the 20 percent who do not farm represent the entire food-purchasing population. Consider Kenya, for example, where it was estimated in 1965 that 88 percent of the economically active population was in agriculture and only 7.8 percent of the "African" (black) population lived in towns of 2,000 or more. Imperfect data suggest that in 1960/61, purchases of maize, the principal starch staple, amounted to not less than 60 percent of all maize consumed (W. O. Jones 1972:44–45, 52, 206–7, 217). Who were the purchasers? Some of them must have been farmers, or at any rate "engaged in agriculture." These would include men and women who worked for wages on farms, some quite large, that were owned by others; and we know that employees of plantations and large field-crop farms were paid their wages in cash, not, as in an earlier period, in kind. Reference has already been made to the trade between farmers and pastoralists and this must account for more of the maize sales.

Unfortunately, little information is available about food purchases and household budgets of rural people. Some bits of information about Central Province, Kenya, suggest that a sizable proportion of the purchasers of maize may in fact be farmers. Central Province is the homeland of the Kikuyu—whose trade with the Masai has already been referred to. Before the British occupation of Kenya, the Kikuyu were known as good farmers from whom caravans could usually secure provisions. This is the area where development of commercial African farming under the Swynnerton

[12]Elkan suggests that the low regard contemporary economists have for description frequently causes the vocabulary to get "in the way of viewing African economies as they really are. . . . We have become so accustomed to thinking of them [industry and manufacture] as being associated with factory buildings enclosing power-driven machinery that . . . we have totally ignored a great deal of industrial activity in African countries which has gone on under our very noses" (Elkan 1976:693).

Plan has been most successful, and the province probably produces slightly more maize than it consumes. Variations in altitude, however, result in microclimatic differentiation of production, and there is a lively trade within the province in locally produced foodstuffs. Farmers in the high country bring to market firewood, charcoal, vegetables, tea, dairy products, and wattle bark; the middle elevations supply coffee, maize, beans, bananas, and sweet potatoes; and the "lowlands" (still above 4,000 feet) sell millet, sorghum, gourds, sisal, and livestock (Jones 1972:213).

A survey of Central Province conducted in 1963/64 by the Kenya Ministry of Economic Planning and Development (1968) gives some notion of food purchases by rural households in four of the five districts of the province. Table 3.1 shows production, sales, purchases, and consumption of maize for households in the sample. In three of the districts, purchases accounted for close to one-fourth or more of total maize consumption.[13]

The principal sources of cash receipts from economic activity for farmers in the province (including Embu District) are shown in Table 3.2. It must be emphasized that these were truly rural households (a separate urban survey was carried out in the province). Nevertheless, an industrial classification of household members aged fifteen years and over showed 20 percent of the men and about 4 percent of the women to be engaged in service activities and about 12 percent of each group engaged in unspecified, nonagricultural activities (Kenya 1968:13).

*Table 3.1.* Maize production, sales, purchases, and apparent consumption, Central Province, Kenya, 1963/64*

| District | Million calories per household | | | | Percent of consumption purchased |
| | Production | Sales | Purchases | Apparent consumption | |
|---|---|---|---|---|---|
| Kiambu | 1.02 | .07 | .43 | 1.39 | *31* |
| Fort Hall | 1.77 | .20 | .49 | 2.06 | *24* |
| Nyeri | 1.62 | .23 | .42 | 1.81 | *23* |
| Meru | 1.82 | .43 | .09 | 1.48 | *6* |

*Based on Kenya, Ministry of Economic Planning and Development, Statistical Division, *Economic Survey of Central Province, 1963/64* (Nairobi, 1968), pp. 22, 30, 53.

[13]Consumption of maize per capita per day averaged 725 calories in Kiambu, 1,025 in Fort Hall, 900 in Nyeri, and 700 in Meru.

Table 3.2. Principal sources of cash receipts from economic activity for farmers in Central Province, Kenya, 1963/64*

|  | Percent |  | Percent |
|---|---|---|---|
| Sale of crops | 23 | Manufacturing |  |
| Coffee | (11) | and crafts | 9 |
| Sale of livestock |  | Trade (net) | 5 |
| and products | 14 | Transport | 2 |
| Milk | (6) | Services | 49 |

*Based on Kenya, Ministry of Economic Planning and Development, Statistical Division, *Economic Survey of Central Province, 1963/64* (Nairobi, 1968), pp. 28, 33. Dividends, interest, rent, remittances from relatives, and other miscellaneous receipts are not included. They equaled 14 percent of receipts from economic activities.

## Productive Activities of Rural Households

The data from the Central Province study suggest two characteristics of African society that may stimulate the market for domestically produced foodstuffs. In the first place, it seems clear that even in an area that produces an overall surplus of a particular foodstuff, some households will have more, others less than they wish to consume. Secondly, the distinction between rural and urban populations does not automatically provide a distinction between farming and nonfarming activities. More than this, a high proportion of persons who report farming as a major occupation may engage in other activities that require comparable amounts of time and yield comparable or greater monetary returns.

Lack of correspondence of the category "rural" with farming and "urban" with nonfarming is most striking in the former Western State of Nigeria. Unlike the Kikuyu of Central Province, who are not by tradition village dwellers and who prefer to live on their farms, the Yoruba farmers of the Western State prefer to live in towns, and in the early 1950s approximately 50 percent of the population was to be found in settlements of 5,000 or more. Household accounts of farmers in the cocoa region of Nigeria that were collected by Robert Galletti and others in 1951/52 (Galetti et al. 1956) illustrate both the diversity of the sources of income of these "cocoa farmers" and the extent to which market purchases contributed to the essential food supply (see Table 3.3). Farming activities accounted for 63 percent of net cash revenues (higher than for Central Prov-

*Table 3.3.* Cash accounts for 187 families of the cocoa-growing regions of Nigeria, 1951/52
(£ *West African per household*)*

| Business and capital account | | | |
| --- | --- | --- | --- |

### Farming

| Receipts from sales | | Expenses | |
| --- | --- | --- | --- |
| Food crops[a] | 14.1 | Wages | 18.3 |
| Tree crops[b] | 111.0 | Seed and other | 1.7 |
| Other crops | 1.0 | Underreported | 5.8 |
| Processed food | 2.5 | | |
| Livestock | 1.1 | | |
| Underreported[c] | 1.3 | | |
| Total farming receipts | 131.0 | Total farming expenses | 25.8 |
| | | Net cash return | 104.2 |

### Trade

| Receipts from sales | | Expenses | |
| --- | --- | --- | --- |
| Prepared food | 14.3 | Purchase of merchandise | 231.0 |
| Imported goods | 12.4 | Other | 0.9 |
| Miscellaneous | 40.7 | Underreported | 10.1 |
| Produce | 211.1 | | |
| Underreported | 1.1 | | |
| Total trade receipts | 279.6 | Total trade expenses | 242.0 |
| | | Net cash return | 37.6 |

### Other resources

| Receipts from sales | | Expenses | |
| --- | --- | --- | --- |
| Home industry[d] | 5.2 | Raw materials | 1.4 |
| Services | 7.4 | Underreported | 0.7 |
| Forest produce | 1.1 | | |
| Remittances and gifts | 9.7 | | |
| Miscellaneous | 3.4 | | |
| Total other receipts | 26.8 | Total other expenses | 2.1 |
| | | Net cash return | 24.7 |

| | |
| --- | --- |
| Net cash income on revenue account | 167.5 |
| Net cash income on capital account, less taxes | 14.4 |
| Balance carried to domestic account | 181.9 |

| Domestic account | | | |
| --- | --- | --- | --- |

### Expenses

| Household | | Ceremonial | 5.4 |
| --- | --- | --- | --- |
| Building | 7.2 | Social | |
| Food | 47.3 | Church and club dues | 1.9 |
| Clothing | 11.9 | Entertainment and gifts | 7.3 |
| Drink, tobacco, kola | 8.0 | Total social expenses | 9.2 |
| Furniture | 0.6 | Miscellaneous | |
| Fuel, light, other | 2.2 | Travel | 1.5 |

*Table 3.3.* (cont.)

| Domestic account | | | |
|---|---|---|---|
| Expenses | | | |
| Education | 6.0 | Litigation | 0.6 |
| Medical | 1.7 | Other | 10.6 |
| | | Total miscellaneous | |
| | | expenses | 12.7 |
| Total household expenses | 84.9 | Total domestic expenses | 112.2 |
| | | Addition to cash balance | |
| | | at end of survey year | 69.7 |

*Based on R. Galletti, K. D. S. Baldwin, and I. O. Dina, *Nigerian Cocoa Farmers: An Economic Survey of Yoruba Cocoa Farming Families* (London, 1956), p. 554.
[a]Starchy staples—yams, manioc, maize, plantains, and rice—accounted for 86 percent of food crop sales.
[b]Cocoa, kola, palm fruit and kernels.
[c]Galletti and his associates attempted to adjust the figures for underreporting that they were convinced had occurred (cf. Galletti et al. 1956: 546–52). They say that the adjustments for underreporting "are not exact estimates of the margins of error but do in our opinion give a fair indication of the errors which have affected the correctness of the accounts and of their relative importance" (Galletti 1956: 552).
[d]Mostly weaving.

ince, Kenya), and trade for 22 percent. Farming expenses accounted for less than 20 percent of farm receipts, whereas those members of the household, mostly women, who were engaged in trade paid out 87 percent of their receipts, primarily to buy stock.

In 1966/67, the Nigerian Institute of Social and Economic Research (NISER), in cooperation with the Food and Agriculture Organization of United Nations (FAO), conducted extensive socioeconomic research in the Kainji Lake area to assist in the resettlement of people displaced by a new dam. One part of this research was a survey by O. A. Oguntoye (1968) of the occupants of an area denominated Old Bussa, about Bussa Town, Kwara State, along the south bank of the Niger River. The population numbered about 3,300 persons of mixed ethnic composition with Bussawa predominating, followed by Yoruba, Hausa, and Nupe in that order, but with more than twelve other groups also represented (Oguntoye 1968:1–2). Interviews with 656 inhabitants are summarized in Table 3.4. Oguntoye (1968:7) comments: "From the figures . . . it could be concluded that farming, petty trading and scattered craft industries from the main occupation in the area. Fishing is given as one of the traditional occupations of the people and it is therefore a surprise that very few among those interviewed are full-time fishermen."

*Table 3.4.* Principal occupations of 656 residents of
Old Bussa*

| Occupation | Number | Percent |
|---|---|---|
| Farmers | 229 | 34.9 |
| Craftsmen[a] | 78 | 11.9 |
| Craftswomen[b] | 54 | 8.2 |
| Fishermen and canoe drivers | 40 | 6.1 |
| Traders | 87 | 13.3 |
| Others[c] | 168 | 25.6 |
| Total | 656 | 100.0 |

*Based on O. A. Oguntoye, "Occupational Survey of
Old Bussa," NISER (Ibadan, October 1968, processed).
[a]Butchers (5), baker (1), dyers (5), grinder (1), blacksmiths (4), weavers (10), carpenters (3), tailors (14),
painters (3), builders and bricklayers (16), barber (1),
bicycle repairer (1), radio repairers (2), watch repairer
(1), mat weavers (5), spinners (6).
[b]Spinners.
[c]Government employees (132), employees of foreign
firms (8), contractors (2), religious leaders (15), astro-
loger (1).

The explanation of this apparent inconsistency between general knowledge and the findings of the interviewers is revealed in Table 3.5, which
lists the other activities of persons classed in Table 3.4 as farmers. The
total number of fishermen and canoe drivers, including those who are also
classified as farmers, now becomes 134, or 20 percent of all persons
interviewed. But the confusion between farmers and nonfarmers extends
farther than this. Oguntoye says (1968:12–13) that "The majority of those
engaged in farming activities can not, in fact, be regarded as farmers;
because farming is not their major occupation. These people who plant
between 0.01–2 acres of land generally have another occupation—mainly
with the federal or state government(s), native administration, or firms. . . .
Collectively, they plant a substantial area (180.5 acres). . . . Thus only 184
out of 393 people [47 percent], who hold and plant land can be regarded as
farmers." As Table 3.5 shows, only seventy-four of these are engaged in
farming exclusively.

Anthony and Johnston provide similar information about the other occupations of the sixty farmers they interviewed in Katsina Province (Table
3.6).

W. H. Beckett has recorded occupational data for Akokoaso, Ghana, in
about 1935 (Table 3.7). All but 2 of the 352 men in the population had an

*Table 3.5.* Other occupations of 223 farmers in
Old Bussa*

| Occupation | Number | Percent |
|---|---|---|
| Craftsmen[a] | 31 | 13.9 |
| Fishermen and canoe drivers | 94 | 42.1 |
| Traders | 11 | 4.9 |
| Laborers | 17 | 7.6 |
| Others[b] | 5 | 2.2 |
| No other occupation | 74 | 33.2 |
| Total | 232[c] | 104.0 |

*See source note, Table 3.4.
[a]Weavers (10), builders and bricklayers (3), barber (1),
dyers (11), tailors (5), wine tapper (1).
[b]Drummers (2), teachers of Arabic (2), hunter (1).
[c]Source says 223 people in the title but detail adds to 232,
perhaps because some farmers have more than one
supplementary occupation. Nor does the source reconcile
the total given here with the total number of farmers
shown in Table 3.4.

occupation of some sort: of these, 100 were employed in occupations other than farming or in addition to farming and 22 had taken employment away from the village. There were 163 who owned their farms independent of any landlord; 51 of these had some other occupation. Of the 322 women, 35 had occupations other than farming, 172 farmed only, and 113 had some occupation other than that of housewife (Beckett 1944:15–17).

In a preliminary report on her study of Batagarawa in Katsina Province, Nigeria, Polly Hill comments on the impossibility of identifying a "subsistence" sector (Hill 1970:147):

Farming is much the most important occupation in Batagarawa. Grains (millet and guinea corn) and groundnuts (mainly for export) are the chief crops, and numerous other crops including beans (*wake*) and native (not commercial) tobacco are also grown. It is nothing but misleading to refer to cash and subsistence "sectors": any crop might happen to be sold by the grower for cash and there are few crops other than tobacco, which are not apt to be self-consumed. Nor does the economist's word "surplus" have any useful meaning: the better-off people aim at buying grain when it is cheap (just because it is cheap) and at selling it when it is dear, the worse-off people are often obliged to do exactly the opposite.

Later in the same chapter, Hill describes the great differences in income and wealth among the households of the village and states that nearly all of

*Table 3.6.* Other occupations of 60
farmers in northern Katsina*

| Occupation | Number | Percent |
|---|---|---|
| Craftsmen[a] | 17 | 28.3 |
| Traders | 9 | 15.0 |
| Laborers | 2 | 3.3 |
| Others[b] | 3 | 5.0 |
| Total | 31 | 51.6 |

*Data from K. R. M. Anthony and
B. F. Johnston, "Field Study of Agricultural Change: Northern Katsina,
Nigeria," Food Research Institute
Study of Economic, Cultural, and
Technical Determinants of Agricultural Change in Tropical Africa, Preliminary Report No. 6 (Stanford,
1968, processed).
[a]Tailors (4), weavers (3), builders (2),
blacksmiths (2), dyers (2), potter (1),
rope maker (1), butcher (1), well
digger (1).
[b]Scribe, Islamic teacher, and herds-
man.

the top 10 percent had "reasonably lucrative non-farming occupations,
mainly craft (notably tailoring) and local trading in grains and
groundnuts."[14] The very poor also sought employment as farm laborers or
collecting and selling firewood, and a few deliberately contracted out of
farming to "attach themselves as 'servants' to the ruling class—a group
which includes courtiers, drummers, flatterers, etc., as well as house and
farm servants." Most of the "failed" farmers depended "on the sale or
manufacture of 'free goods' (grass, twigs, cornstalks, palm leaves, wood,
etc. from which mats, rope, thatches, beds, etc., are made for sale in
Batagarawa) or on farm or general labouring" (Hill 1970:152–53).

Few attempts have been made to record the hours spent by rural people
on activities other than farming, and frequently records of farming activity
include only time spent in the field. John Cleave's examination of farm
surveys in English-speaking countries on both sides of tropical Africa

[14]In a careful study of rural incomes in Central Province, Kenya, in 1963, T. Kmietowicz
and P. Webley found mean incomes of 1,238s., modal income of 583s., and standard
deviation of 1,208s. The Gini coefficient is 0.43, slightly greater than for the total population
of the United Kingdom and about the same as for incomes of residents, mostly Asians and
Europeans, who paid income tax at that time (Kmietowicz and Webley 1975).

shows that the time actually spent in farming proper (by adult males) ranges from about 530 to 2,135 hours per year, but with all areas but one reporting less than 1,700 hours (Table 3.8).

It seems extremely unlikely that the balance of the daylight hours is spent in complete idleness; fragmentary evidence suggests that some of the time that is unaccounted for may be employed to produce additional income. M. P. Collinson presents data based on D. Pudsey's studies of twelve farms in Toro, Uganda, that account for 7.1 to 9.6 hours per day, assuming 300 working days in the year. They show nonfarm activities to account for from 3.5 to 8.7 hours a day (Collinson, 1972:36). Many of these activities, although productive, yielded no money income, and Pudsey's data do not distinguish those that did. Judith Heyer, however, in an unpublished paper cited by Collinson, reports that fourteen farmers in Machakos, Kenya, "used 37 percent of available time over the year on crop work and a further 26 percent on nonspecific work directly associated with agriculture. Other work included beer brewing, crop processing, marketing, craft work, cattle herding, and contract services" (Collinson 1972:34).

In a study of employment in three villages in the Zaria area of northern Nigeria in 1966–67, David Norman found the sixty-nine men in farming families working from 254 to 257 days a year, but only spending a little

Table 3.7. Other occupations of farmers in Akokoaso, Ghana*

| Occupation | Number | Percent of all farmers |
|---|---|---|
| Craftsmen[a] | 23 | 4.9 |
| Tappers[b] | 18 | 3.9 |
| Hunters | 15 | 3.2 |
| Traders[c] | 7 | 1.5 |
| Doctors | 3 | 0.6 |
| Forest guards | 1 | — |
| Total | 67 | 14.3[d] |

*Data from W. H. Beckett, *Akokoaso: A Survey of a Gold Coast Village* (London, 1944), p. 16.
[a]Potters (8, all women), masons (5), shoemakers (3), sawyers (2), blacksmiths (2), carpenter (1), others (2).
[b]Of palm wine.
[c]Including one cocoa buyer.
[d]Detail does not add to total because of rounding.

Table 3.8. Time spent in agricultural work by adult males (by number of surveys reporting the indicated hours per year)*

| Supply area | Hours per year[a] | | | | | | | | |
|---|---|---|---|---|---|---|---|---|---|
| | 500 to 700 | 700 to 900 | 900 to 1,100 | 1,100 to 1,300 | 1,300 to 1,500 | 1,500 to 1,700 | 1,700 to 1,900 | 1,900 to 2,100 | 2,100 to 2,300 |
| Nigeria | | | | | | | | | |
|   Cocoa area | | | 3 | 1 | | | | | |
|   Southwest | | 1 | | | 1 | 3 | 1 | | 1 |
|   Northern | 1 | 1 | | | | | | | |
|   North Central | 4 | | 1 | | | | | | |
| Ghana | | | | | | | | | |
|   Akokoaso | | | | | | 1 | | | |
|   Battor | 1 | | | | | | | | |
| Gambia | | | | | | | | | |
|   Genieri | 1 | | | | | | | | |
| Rhodesia | | | | | | | | | |
|   Chitowa | | | 1 | | | | | | |
| Uganda | | | | | | | | | |
|   Western | | | 2 | 1 | | | | | |
|   Northern | 1 | 1 | 2 | | | | | | |

*Data from J. H. Cleave, African Farmers: Labor Use in the Development of Smallholder Agriculture (New York, 1974), pp. 32–33.
[a]Does not include time spent in walking to and from the fields where this is specified in sources used by Cleave.

over half of this time in farming. The time spent in other occupations was allocated as shown in Table 3.9.

Roger de Smet provides some information from Zaïre, based on a time-budget study of eleven Zande households in Mangara District of the southern Uélé from July 1958 to April 1961. This is a forested area where palm fruit was the principal cash crop and bananas the principal food crop. Hunting, fishing, and gathering were also important sources of foodstuffs. Table 3.10 shows how the adults of these households employed their time. Employment shows strong seasonal variation, with women's nonworking hours varying from 13 percent of the day in March, when gathering activities (fishing?) are at their peak, to 40 percent in October, when all activities except housekeeping are at a minimum. Men are busiest from February through June, when nonwork accounts for less than a third of their available time, and least busy in August and October, when more than half of their time is free. Seasonal variation in the time devoted to farming, manufacturing, building, hunting and gathering, and social activities is much greater. Construction, for example, takes up 17.5 percent of the men's time from July through November, but less than 5 percent in any of the other months. And housekeeping activities of women show a complementarity with time spent on the farms.

*Table 3.9.* Time spent in occupations other than farming by villagers in northern Nigeria, 1966/67*

|  | Percent of time | |
| --- | --- | --- |
| Occupation | Small farms[a] | Large farms |
| Manufacturing[b] | 22.6 | 20.3 |
| Traditional services[c] | 41.4 | 39.0 |
| Modern services[d] | 0.4 | 6.0 |
| Trading | 35.6 | 34.7 |

*Data from D. W. Norman, "Economic Analysis of Agricultural Production and Labor Utilization among the Hausa in the North of Nigeria," Michigan State University African Rural Employment Study Paper No. 4 (Ann Arbor, January 1973), p. 1.
[a]Less than 1½ acres per household member.
[b]Including blacksmiths, tailors, carpenters, spinners, leather workers, potters, cigarette makers, mat weavers, and sugar makers.
[c]Including builders, thatchers, grass and firewood cutters, bakers, butchers, hunters, washermen, beggars, and Koran teachers.
[d]Including commission agents, messengers, laborers, night watchmen, bicycle mechanics, and buying agents.

*Table 3.10.* Use of time by adults of Zande households in Zaïre, July
1958–April 1961*

| | Percent of "available time"[a] | |
|---|---|---|
| | Men | Women |
| Farming | 20.3 | 32.4 |
| Hunting, fishing, gathering | 14.4 | 9.2 |
| Housekeeping and agricultural processing | 3.7 | 27.8 |
| Manufacturing | 4.3 | 0.7 |
| Building | 8.1 | 1.6 |
| Service activities[b] | 11.2 | 6.7 |
| Nonwork[c] | 38.0 | 21.9 |

*Data from R. E. de Smet, *"Une Enquête budget-temps dan les Uélé
République du Zaïre,"* in *Etudes de geographie tropicale offertes à Pierre
Gourou* (Paris, 1972), p. 294.
[a]Defined as daylight hours, i.e., 6:00 A.M. to 6:00 P.M., less time lost due to
illness (4.3 percent) and time spent visiting relatives (1.6 percent for men
and 3 percent for women).
[b]Administration, adjudication, trading, and social activities.
[c]Leisure, eating, personal hygiene.

Among the Gouro of central Ivory Coast, on the other hand, all the many
crafts once produced by rural people are now supplied by migrant
craftsmen. Despite increasing competition from printed cloth, however,
the Gouro continue to manufacture the traditional brightly colored
loincloths (pagnes) and the basket trays used for winnowing, being reluctant to rely on the market for these items (Meillassoux 1964:189).

In a summary of a Michigan State University five-year study of rural
employment in tropical Africa, the authors conclude that "nonfarm activity
in the rural areas provides a source of primary or secondary employment
for 30 to 50 percent of the rural male labor force in tropical Africa," and
they estimate that trading and manufacturing account for more than 70
percent of employment, presumably of men, in the rural nonfarm sector
(Byerlee et al. 1977:22, 24).

How far the mixtures of productive activities revealed by these studies
are characteristic of African rural households cannot be determined with
the information that is now available, but there is good reason to believe
they may be widespread. Part of the problem is that most studies of rural
communities (they are not many) have concentrated on farming activities,
often even excluding the first stage of processing that goes on in the
farmyard—drying coffee, retting and pounding manioc or making it into
*gari,* grinding or pounding maize, millet, or rice—and have completely

ignored other productive activities. Part of the problem, too, is that many farm residents find part-time employment off the farm; a number of surveys report large numbers of the able-bodied men were away from their homes in wage employment during much of the year.

## The Domestic Demand for Food Crops

African newspapers tend to carry more stories about insufficiencies of foodstuffs than they do about insufficient demand, perhaps reflecting their urban readership, but if farm production is to increase, farmers must be assured of markets for their harvests. A review of what is known about the present and prospective consumer demand for purchased foodstuffs strongly suggests that domestic market opportunities are large and will grow. It also confirms that the largest potential market opportunity in the near term may be in supplying the households of producers of export crops.

### Rural Demand for Food

The Galletti study mentioned earlier also provides detailed information on food purchases that permits calculation of the proportion of starchy staples, in calorie terms, that the western Nigerian farmers obtained by purchase—approximately 60 percent, with yams and manioc and their products accounting for the bulk of it. Galletti's findings are confirmed by the Rural Consumption Enquiry conducted by the Nigerian Federal Office of Statistics about ten years later. This survey found that in seven western Nigerian villages, purchases accounted for 58 percent of all calories from staples with a range from 46 to 77 percent. These are high figures and are probably equaled in very few other parts of tropical Africa. For seven eastern Nigerian communities, for example, the Office of Statistics figures average only 28 percent (Jones 1972:60–61, 64).

A study of the expenditures of cocoa-producing families in the Oda-Swedru-Asamankese District of Ghana in 1955/56 showed these families to be spending 1,095 shillings per household on food, as compared with 925 shillings per family in the Galletti study, carried out four years earlier. The comparable percentages of total consumption expenses were 50 for the Ghana farmers, 59 for the Nigerian ones. The comparability of these figures is intriguing, but unfortunately the Ghanaian figures on food expenditure were not published in sufficient detail to permit calculation of percentage of staple foodstuffs purchased. The Ghanaian farm households obtained almost all of their cash income from farming (Ghana 1958).

Farmers in Bongouanou District in the Ivory Coast, also an area of tree-crop production (both cocoa and coffee), represented a contrasting situation. According to a sample survey in 1955/56, farm households then were still producing virtually all of their food requirements—89 percent in terms of value and a still higher proportion of their calorie intake. This was in spite of an increase in the population of the district—from 15,000-20,000 in 1910 to nearly 65,000 in 1955—and expansion of coffee and cocoa production to the extent that those crops required more labor than the food crops—mainly yams and plantains, which give high yields per unit of land and labor (Côte d'Ivoire 1958:16). Since that time there has been a large increase in the production of rice by migrant farmers on land too wet for tree crops. An unknown share of this production is purchased by local cocoa and coffee farmers.

In 1961/62, Ghana carried out a carefully planned study over a three-month period of 3,000 households randomly selected throughout the country. The results of this survey were never fully analyzed, but a sample drawn from the total provides indications of the amount of food purchases by rural households. For all rural Ghana, the survey indicates that from 52 to 54 percent of all food, measured in terms of value, was purchased. The percent of food purchased varied regionally from a low of 24 to 29 percent in the Northern Region to about 70 percent in the Western Region (Ord et al. 1964:105).

Sukumaland in Tanzania is another area where greatly increased production of an export crop—cotton—led to increasing reliance on purchases of staple foods. This caused concern to British colonial administrators, who attempted over a period of years to achieve local self-sufficiency in foodstuffs. In 1950, for example, the report of the Department of Agriculture on the Southern Province complained that "over recent years the peasant has become increasingly money conscious and has been developing the idea that if you have money you cannot be short of food" (Jones 1960:119). Fortunately for the Sukuma, the British were not successful in their attempts to inhibit the trade in staple foodstuffs, and in 1967 Anthony and Uchendu found farmers in Geita District, which had rather recently been opened to crop agriculture with cotton as the principal cash crop, to be supplying substantial quantities of manioc to the central areas of Sukumaland, with some district farmers specializing in the production of manioc for sale (Anthony and Uchendu 1974:70):

> The recent expansion of cassava production for sale to other areas in Sukumaland is another success story. It has resulted from the response of farmers to the relative prices for the commodities they produce. At the time of the Food Research Institute Survey, there were some farmers who derived most of their cash income from cassava. Despite the pres-

sure on farmers to produce more cotton, this trend is likely to continue. The trade in cassava is handled by small middlemen and the marketing organization that has been developed is probably more efficient than that of the co-operative societies.

Demand for manioc in the older cotton-growing areas of Sukumaland resulted from the increasing value of farmland there as population rose and from the difficulty of protecting the crop against livestock damage (Anthony and Uchendu 1974:18).

Kenneth Shapiro also reports that in his survey of seventy-six farm households in Geita District in 1969–71, although cotton was by far the most important cash crop, "All food crops also serve as cash crops, but in a minor way for most farmers. However, a few farmers rely on cassava or rice or both for most of their cash income" (Shapiro 1973: 16).

Cleave's study of farm accounts concludes that, although farms included in these studies tended to be highly commercialized in terms of the proportion of total production that was sold, "In most areas ... farmers are continuing to provide all or most of their basic food requirements in addition to growing crops for sale. Of the farmers who are dependent on the market for their food, the Yoruba cocoa growers and the oil palm farmers of Calabar have the highest real income of any of those studies ... and the tea growers of ... Toro [Uganda] ... are extreme examples of a commitment to a crop for sale. Even so, on the average farm in all these areas at least half of a subsistence diet is produced" (Cleave 1974:28–29).

On the other hand, Cleave provides interesting detail from a study of Toro District farmers made by Pudsey that illustrates how staple food purchases can vary with the source and size of cash income (Pudsey 1967). In Kahangi, thirty farmers reported that on the average, 11 percent of gross farm income came from sales, and twenty of the farmers for whom data on staple food production was reported were judged to be "completely self-sufficient in basic calories." In Kyarosozi, twenty-six farms were surveyed and reported sales amounting to 63 percent of gross farm income. The nine farmers who grew tea and brewed beer for sale but had no outside income produced "the larger part of their staple needs." The other seventeen farmers had regular off-farm income and were "clearly dependent on the market even for staples." Calorie values per head per day of staple foods consumed from own production by each of the three groups are shown in Table 3.11.

One other study, this too of Uganda, carries the search for information about the market for domestically produced food a bit further. This is William Mackenzie's comparison of estimated production of staple

*Table 3.11.* Calorie values per head per day of staple foods consumed from own production by three groups of Toro, Uganda farmers *

| | | Kyarosozi | |
|---|---|---|---|
| Foodstuff | Kahangi self-sufficient farmers | Farmers without outside income | Farmers with outside income |
| Bananas | 930 | 375 | 250 |
| Beans | 265 | 200 | 230 |
| Millet | 275 | 285 | 235 |
| Sweet potatoes | 365 | 510 | 355 |
| Total | 1,835 | 1,370 | 1,070 |

*Data from J. H. Cleave, *African Farmers: Labor Use in the Development of Smallholder Agriculture* (New York, 1974), p. 217.

foodstuffs and estimated calorie requirements by district from 1955 to 1967. It demonstrates rather clearly that there must have been a fairly strong interdistrict trade in staple foodstuffs.[15] Mackenzie's calculation of apparent calories available per head per day from local production of staples in 1964 (Table 3.12) shows a range from 3,464 in Bunyoro to 643 in Karamoja. East and West Mengo and Busoga embrace the most urbanized parts of Uganda, but they are not very urbanized and they are not the only districts that appear to show local supplies to be inadequate for requirements.[16] According to Table 2.1, 86 percent of the economically active population of Uganda is engaged in farming. Even in Zaïre and Zambia, with their large mineral production, the figures are 80 and 73 percent.

Apparently, estimates of the percent of the labor force engaged in agriculture are not a very good indication of the probable market demand for domestically produced foodstuffs. Such figures can only be crude approximations, and they fail to indicate the extent to which the rural population supplies other goods and services than peculiarly agricultural ones, and the extent to which its agricultural activities are directed toward production for

[15] Berndt Schubert, who conducted studies of marketing in Uganda in the 1960s, states in a letter of November 27, 1973, that "the fact that there is an interdistrict and interregional food exchange [in Uganda] has . . . long been observed by the agricultural administration." He further says that this is reported in the annual reports of provincial and district agricultural offices since 1945. Nevertheless, up to the time when Mackenzie wrote, gross domestic product in Uganda was calculated on the assumption that each province was self-sufficient in staple foodstuffs.

[16] Vali Jamal (1976) has pointed out that a considerable part of the shortfall in calories in Mengo may have been made good by imports.

Table 3.12. Apparent number of calories available per
head per day from local production of staples in Uganda
districts, 1964*

| District | Calories | District | Calories |
|---|---|---|---|
| Bunyoro | 3,464 | Bugisu-Sebi | 2,002 |
| Teso | 3,078 | Busoga | 1,990 |
| Acholi | 3,050 | Bukedi | 1,847 |
| Lango | 2,874 | Masaka | 1,725 |
| West Nile-Madi | 2,358 | East Mengo | 1,698 |
| Toro | 2,186 | Ankole | 1,413 |
| Kigezi | 2,093 | West Mengo | 1,300 |
| Mubende | 2,020 | Karamoja[a] | 643 |

*Data from William MacKenzie, "The Use of Aggregate
Data in the Study of Agricultural Change: A Case Study
of Uganda, 1955–67," *Food Research Institute Studies
10,* (1971), p. 33. Commodities included millet, sorghum, maize, manioc, sweet potatoes, bananas, peanuts,
beans, peas, and sesame.
[a]The inhabitants of Karamoja (the Karamajong) numbered only 172,000 in 1962 and their diet relied heavily
on animal products (Thrower 1966: map).

market sales. The second column of Table 2.1, which shows agricultural
production as a percentage of gross domestic product, may be a somewhat
better guide, but it, too, glosses over the extent of intra-agricultural trade,
which of itself is cost-reducing. Furthermore, most censuses of manufactures in the African countries are underestimates because they have recorded the output only of firms with at least five employees (sometimes
ten), and consequently fail to include values of craft manufactures and
services that may be large.[17]

## Import Substitution

A very considerable part of the total value of foodstuffs that are purchased by African consumers is made up of imported commodities. In a
few but increasing number of instances these are carbohydrates, like rice

[17]A. D. Goddard's study of industry in Zaria provides a case in point. Goddard says
(1970:170): "Manufacturing industry in Zaria falls into two distinct categories.... There are
a few relatively large firms which involve large amounts of foreign or government capital,
using advanced technologies, experienced management and a skilled workforce... [and]
numerous indigenous crafts and small workshops.... The former tend to dominate many
peoples' thoughts... but... they have only small workforces in relation to the capital
invested.... The employment... [the small industries] provide far exceeds that provided by
the big firms."

and sugar, that are or could be grown in the importing country or that compete directly with locally grown starchy staples, as does wheat. But the major food imports are relatively highly valued products like canned or dried meat and fish, processed milk, a wide range of canned goods, and other delicacies. For many of these products, the value added by processors and merchants is likely to be greater than that added by the farmer or fisherman who supplied the original products. In general, the opportunities for increasing the role of domestic agricultural output by substituting it for imported foodstuffs are few. Any sizable increase in agricultural production for domestic sale must look for its market among consumers who already have or will substitute purchased foodstuffs for those they grow themselves and who will increase the amount they spend on food purchases as their incomes grow.

It has already been argued that market demand will grow with urbanization, expansion of manufacturing, utilities, transport, and public services, and with increased specialization in agriculture. It is necessary now to consider the little that is known about how expenditures on food change with income—the income elasticity of demand.

## Elasticity of Demand for Staple Foods

One of the earlier attempts to estimate income elasticities of demand for food in tropical Africa was that by Hiromitsu Kaneda and B. F. Johnston in 1961. On the basis of twenty-three household expenditure studies in fourteen cities in eight countries, they compared total household expenditures with expenditures on all food and on starchy staples. The figures appear to show the expected decline in percent of expenditures going for food as total expenditures rise, i.e., elasticity less than 1.0, but the observations are strongly clustered and could not reasonably be used as a basis for statistical estimates. When the data permitted, Kaneda and Johnston also examined the elasticity relationship within a particular study, with less orthodox results. A study of household expenditures in Kumasi, Ghana, appears to show the percent of expenditures for food first increasing with total expenditures, then decreasing, but this may be caused by varying family size (Kaneda and Johnston 1961:237).[18] The income elasticity of demand for the starchy staples seems to be markedly lower than for all food, as would be expected.

Three Ghana studies examined in detail by T. T. Poleman—Accra, Kumasi, and Sekondi-Takoradi—show per capita food outlay to be increas-

[18]A somewhat similar phenomenon has been reported in Akuse, Ghana, with the demand for dried manioc, usually considered to be an inferior good (Jones 1959:201).

ing almost equally with family size, and the percent spent on food declining very slowly with increased total expenditures (Poleman 1961:145).[19] These studies also show purchases of all starchy staples to be a remarkably constant proportion of total food purchases, whether the household spent as little as 125 shillings a month or as much as 650 shillings (Poleman 1961:155). Within the set of starchy staples, however, there was a tendency to shift from those costing less per calorie to those costing more— presumably from less to more preferred staples—with rice and wheat bread always in the favored group, and fresh manioc sometimes.[20] In Accra, maize is the most important staple by far, but its income elasticity is clearly negative (Poleman:156). Surprisingly, Poleman found that these studies did not provide clear evidence of a positive income elasticity of demand for fresh meat, although there are indications that income elasticity of demand for fresh fish is greater than zero.

H. W. Ord and his associates calculated income elasticities of demand from the data used by Poleman in these and other similar studies in Ghana and from the National Household Survey of 1961/62. Ord agrees that the proportion of total expenditures devoted to food "was remarkably constant over a wide range of total expenditure levels within lower income samples; and for families in the lowest expenditure 'brackets' it might appear that Engel's Law was invalid." His calculation of income elasticity coefficients over a wide range of incomes is consistent with Engel, however, and Ord suggests that apparently stable proportions in the poorer household may result in part from underreporting, primarily by the women, of expenditures on nonfood items and of income. Income elasticity coefficients for all food, calculated from data in the National Household Survey on a per capita basis, were .76 to .78 in urban areas and .78 in rural areas, for a

---

[19]H. S. Houthakker (1957) calculated expenditure elasticity of demand from these three studies and a similar one in Akuse, Ghana, and also adjusted the estimates crudely for family size:

| Town | Unadjusted elasticity | Adjusted elasticity |
| --- | --- | --- |
| Accra | .95 | .84 |
| Kumasi | .95 | .82 |
| Sekondi-Takoradi | .82 | .65 |
| Akuse | .87 | .79 |

[20]It may be questioned whether fresh manioc eaten by an urban population should be regarded as a starchy staple or a vegetable.

national figure of .77. The coefficients for local food were .58 or .59 in urban areas and .68 or .72 in rural areas (Ord et al. 1964:47, 49, 106).[21] Coefficients on a per capita basis derived from the individual surveys are shown in Table 3.13.

Various studies have been made of household expenditures in some of the major cities of Nigeria—notably Kaduna and Zaria, Ibadan, Lagos, Onitsha, and Enugu—but attempts to calculate elasticity coefficients from them are not convincing. In 1966 the Federal Office of Statistics derived income elasticities for selected groups of foodstuffs. They ranged from .38 to .52 for "staples," from .38 to .69 for meat, fish, and eggs; and from .40 to .62 for total food (Okurume 1969:33). S. O. Adamu (1966) has criticized these estimates severely because they are based on simple linear regressions that in fact did not fit the data very well.[22]

For Kenya and Uganda, B. F. Massell and Andrew Parnes provide estimates of expenditure elasticities for both rural and urban populations (Table 3.14). For the rural areas, "expenditures" were taken to include the value of own production consumed. For Nairobi, they are cash expenditures only. It is noteworthy that elasticities are uniformly much higher for rural than for urban consumers, but because of the statistical unreliability of the Nairobi coefficients, not too much should be made of this difference.

Table 3.13. Per capita income elasticities of demand for food in Ghana*

|  | Local food | All food |
|---|---|---|
| Urban |  |  |
| Accra, 1953 | .81 | .85 |
| Kumasi, 1955 | .52 | .64 |
| Sekondi-Takoradi, 1955 | .45 | .63 |
| Rural |  |  |
| Oda-Swedru | .50 | .59 |
| Ashanti, 1956–57 | .31 | .39 |

*Data from H. W. Ord et al., "Projected Level of Demand, Supply, and Imports of Agricultural Products in 1965, 1970, and 1975," USDA, ERS/FAS and University of Edinburgh (March 1964), p. 49. The considerably higher values from the National Household Survey are not explained. The survey did cover a wider cross section of income and took fuller account of own consumption.

[21]The variations result from including or excluding own consumption.

[22]Detailed elasticity coefficients for individual foodstuffs by regions presented by FAO (1966:398) and reproduced by E. H. Whetham (1972:26) are "assumed" values.

*Table 3.14.* Estimated expenditure elasticities in rural Kenya and Uganda, and in Nairobi, 1963–66*

| Commodity | Western Uganda | Central Province Kenya | Nairobi |
|---|---|---|---|
| Milk | 2.94[a] | 2.34[a] | .48[b] |
| Meat | 1.30[a] | 1.20[a] | .49[b] |
| Cereals | .97[b] | .59[a] | .19 |
| Sugar | .96[b] | 1.06[a] | .23 |
| Pulses | .76[a] | .79[a] | −.06 |
| Vegetables and fruit[c] | .59 | .79[a] | .36 |
| Roots and tubers | −.18 | .53 | — |

*Data from B. F. Massell and Andrew Parnes, "Estimation of Expenditure Elasticity from a Sample of Rural Households in Uganda," *Bulletin of the Oxford University Institute of Economics of Statistics,* 31 (1969), pp. 326–27.
[a]Significant at the 1 percent level.
[b]Significant at the 5 percent level.
[c]Includes bananas.

Massell and Judith Heyer also provide estimates of the expenditure elasticity of demand for all food in Nairobi that range from .39 to .48, depending on the functional form used in calculating them (Massell and Heyer 1969:229).

In Sierra Leone, Robert King has calculated the income elasticity of demand for food by the rural population on the basis of a national sample of 203 households over a period of fourteen months (Table 3.15). Elasticities shown were calculated from a modified ratio semilog inverse function that was chosen over a log-log inverse function because it conforms exactly to "the criterion of additivity of marginal propensities to consume" (King 1977:79). Average elasticities for the entire sample are much higher for cereals and root crops, fruits and vegetables, and palm oil when calculated from a log-log function, but they are about the same for the other commodities and for all food (King 1977:77).

Donald Snyder provides estimates of cash expenditure elasticities of demand for food by 294 households in the Western Area of Sierra Leone in 1967/68. About 90 percent of the households were located in Freetown or its environs. Elasticities calculated from a double-log function were: for all food, 1.15; for fats and oils, .90; for rice, .03; for "cereals," apparently cereals other than rice, .81; for meat, 1.90; for fish, .54; and for condiments, which in Freetown include peanuts and other things used to flavor the "soup," .99 (Snyder 1971: 49–50, 104). Snyder does not comment on the extremely low coefficient for rice. It does not seem consistent with the

*Table 3.15.* Expenditure elasticities of demand of rural populations in Sierra Leone, by income class, 1974/75*

| | Income class—decile[a] | | | | | |
|---|---|---|---|---|---|---|
| Commodity | 1 | 2 & 3 | 4 & 5 | 6 & 7 | 8 & 9 | 10 |
| Bread | 1.26 | .81 | 1.23 | .34 | .26 | 0 |
| Rice | 1.16 | .93 | 1.00 | .89 | 1.00 | 1.01 |
| Cereals and roots | 1.81 | 1.56 | 1.36 | .72 | .51 | .43 |
| Fruits and vegetables | .84 | 1.37 | .80 | .72 | .78 | 1.04 |
| Palm oil | 2.07 | 1.42 | 1.20 | .99 | 1.11 | .80 |
| Livestock products | −1.17 | − .33 | 1.11 | 1.56 | 1.75 | 2.55 |
| Fish | 1.67 | 1.34 | .75 | .90 | .68 | .54 |
| Processed food | .35 | .54 | .99 | .45 | 1.04 | .83 |
| All food | 1.22 | 1.05 | .99 | .87 | .91 | .90 |

*Data from Robert P. King, ''An Analysis of Rural Consumption Patterns in Sierra Leone and Their Employment and Growth Effects,'' Njala University College and Michigan State University African Rural Economy Program, Working Paper No. 21 (East Lansing, March 1977), p. 82.

[a]''Income classes'' are obtained by ranking households by value of per capita consumption, including own consumption.

high value for other cereals, which would be expected to cause them to displace rice as incomes rise.

In multistaple economies like many of the western and central African ones the price elasticity of demand for individual staples can be expected to be high, i.e., a small variation in relative prices will cause a shift from one staple to another. This is important in two regards. On the one hand it means that the risk of serious hunger is much reduced, for consumers can, without difficulty, substitute a less preferred staple that is in good supply for one that is liked better but not plentiful.[23] It is also important from the standpoint of farm production because farmers can increase the amount they offer on the market without this leading to a large decline in prices.[24]

## Elasticity of Demand for Purchased Food

Income elasticities of demand for food appear to be high in most of the studies that have been made in tropical Africa. It is to be expected that the

[23]On the differences between monostaple and multistaple food economies, see Jones (1972:53–54). In fact, monostaple consumers will also shift to a less preferred food when relative prices change drastically.

[24]Sometimes both production and consumption effects can combine to cause farmers to shift from production of a preferred staple for their own consumption to production of it for sale, while at the same time growing a less preferred crop for their own consumption. This appears to happen frequently with yams and manioc in southern Nigeria. It also occurs seasonally (cf. Oluwasanmi et al. 1966).

income elasticity of food *purchases* will be higher than that for all food in households that are in the process of shifting from primary dependence on own supplies to primary dependence on the market. This will probably not be so when cash income, and cash expenditures, are at a very low level. Then almost all food is home-produced and there is no strong sense of need to purchase more, especially not staples. Added income is almost certain to be spent primarily for products that are not or cannot be produced in the rural household—enamelware, for example, or gaily printed textiles—with only trivial amounts for exotic food delicacies. (If bottled soft drinks are foods, and they are, they must be considered as an exception.) Education, too, is likely to be high on the list of priorities. But as more and more of such goods and services are secured, interest will turn to replacing own production with purchased food. For a time, then, the rate of increase in food purchases will be determined by the rate of substitution of purchased for home-grown foodstuffs as well as by the rate of increase in income. This may help to explain why Houthakker felt that he had found some support for the suggestion ''that the elasticity for food with respect to total expenditure might be higher for the countries and time periods with lower average total expenditures, though the evidence is equivocal (Houthakker 1957:547). Adamu presents data for Enugu, Kaduna, and Ibadan that are completely consistent with Houthakker's tentative hypothesis (Table 3.16). Income elasticity of demand for all food and for various classes of food in each of these cities rose with income. To the extent that this proposition is true, estimates of income elasticity of demand for food based on accounts of households that grow some food for their own consumption are apt to be more or less serious underestimates of the impact of increased money income on the market for domestically produced foodstuffs.

R. D. Stevens (1965) has examined the interrelationships among the various determinants of the demand for food in developing countries. He uses six measures of income elasticity at different points in the marketing systems. In terms of the actual and potential market demand for farm crops, the relevant measure is the income elasticity of demand for food at the farm gate. It is determined by the wholesaler's margin, the consumer's income elasticity of demand, and the rate at which consumers replace own production with purchased food as incomes rise.[25] As structural transfor-

---

[25]In symbols, the income elasticity of demand for food at the farm gate (what Stevens calls demand at wholesale) is:

$$e_P = e_R + e_U \tag{1}$$

where e = income elasticity of demand and subscripts indicate the particular variable that is to be related to income: P = food for wholesale; R = food purchased at retail; U = ratio of

*Table 3.16.* Income elasticities of demand for food at three income levels in Enugu, Kaduna, and Ibadan*

| City | Income (*shillings per week*) | Staples | Meat, fish, eggs | Fats and oils | All food |
|---|---|---|---|---|---|
| **Enugu (1961/62)** | | | | | |
| | 50 | .36 | .53 | .49 | — |
| Average | 99.56 | .52 | .69 | .66 | .62 |
| | 150 | .62 | .77 | .74 | — |
| **Kaduna (1962/63)** | | | | | |
| | 80 | .31 | .31 | .28 | .33 |
| Average | 137.99 | .44 | .45 | .40 | .48 |
| | 190 | .52 | .51 | .47 | .54 |
| **Ibadan (1961/62)** | | | | | |
| | 25 | .28 | .42 | .65 | .38 |
| Average | 55.50 | .49 | .64 | .81 | .60 |
| | 100 | .61 | .75 | .88 | .71 |

*Data from S. O. Adamu, "Expenditure Elasticities of Demand for Household Consumer Goods," *Nigerian Journal of Economic and Social Studies* 8 (1966), 481–90.

mation takes place, both consumer elasticity of demand for all food and the rate at which purchased food is substituted for home-produced food can be expected to rise, sometimes causing the elasticity of demand at the farm gate to be greater than unity.[26] Stevens says, "This means that the demand for Food for Wholesale appears likely to grow about as rapidly as per capita income, particularly during the early phases of development" (Stevens 1965:42).

The elasticity of market demand for food at the farm gate may be further increased if the ratio of wholesale to retail value rises, i.e., if marketing costs shrink.[27] They are likely to do so in the early stages of structural

---

wholesale sales to retail sales, i.e., the markup or margin. All figures are in terms of market value.

$$e_R = e_A + e_W \tag{2}$$

where A = all food consumed, W = ratio of food purchased to all food consumed, so that:

$$e_P = e_A + e_W + e_U.$$

This relationship is also discussed by King (1977).

[26]Stevens's formulation suggests that the proportion of food that is purchased can only rise as income does. In fact, it can rise quite independently of income, thus causing market demand to increase even in the absence of a rise in total income. On the other hand, a rise in the proportion of food purchased may be taken as inevitably producing a rise in consumer satisfactions, i.e., in total income, because of the greater freedom of choice that purchase permits.

[27]Marketing margins, too, can change independently of income.

transformation as transport, communications, and other marketing facilities are improved. The volume of demand increases and sources of supply become more specialized. Hunter stresses the interaction between market volume and marketing efficiency (Hunter 1973:247):

> One of the great difficulties in the earliest stages of agricultural de-velopment is the absence of a private commercial sector which can take care of the buying and selling and transport of farm outputs and inputs. This absence is entirely natural. Where farmers have very low purchas-ing power, low productivity, and tiny surpluses to be collected from relatively inaccessible farms, the traders who do exist need very high margins. Often, these margins are attacked as exploitation, but a cooperative or marketing board will also discover to its cost that high costs and risks mean high margins. At a late stage, with better roads, storage, high production, and merchant competition, these difficulties will largely vanish.[28]

Stevens uses arithmetic examples relating changes in marketing margins to changes in income elasticity of demand for food at the farm gate to show that a 5 percent decrease in a marketing margin of 30 percent can raise the farm gate elasticity 38 percent relative to consumer elasticity when income increases 20 percent; similarly, a 5 percent increase in the margin can reduce farm gate elasticity by 41 percent relative to consumer elasticity (Stevens 1965:40). He expects that "considerable reduction in the market-ing margin may be expected in some regions and for some crops where very inefficient marketing systems are used. . . . Major increases in the marketing margin do not appear likely until the later stages of development when consumers begin to demand greatly increased amounts of services with their foods" (Stevens 1965:44).

Examination in detail of the true nature of economic activities in rural areas, and of the potential demand for purchased food in these areas, leads to a moderately optimistic view of the growth of domestic demand for farm products as the members of the society become increasingly accustomed to obtaining their requirements and disposing of their production through market channels. The transition from subsistence to market economy will provide rapidly growing cash outlets for those farmers who seize them; it will also permit them to realize the economies of specialized production.

---

[28]Hunter discusses the danger of unwise government intervention in development of more effective marketing systems, warning that "the tendency to fly straight to imposed coopera-tives or major marketing boards needs far more caution; in too many cases the costs of those organizations and the low prices they give to farmers may be economically less attractive, especially to farmers, than even the merchant service. It may well be better to supply infrastructure and to license traders with adequate, but not extortionate margins" (Hunter 1973:247).

To farmers who produce for market, it will supply both the means and the incentive to employ more conventional purchased inputs such as fertilizers, plows, and pesticides when these can be shown to yield more than they cost.

## The Domestic Marketing System for Staple Foodstuffs

In the near future, the purchasing power of African producers of food crops may depend more on the expansion of food purchases by residents of rural areas than on the increased purchasing power of town dwellers employed in manufacturing, transportation, utilities, services, and other nonagricultural activities. African country people have always exchanged food and other goods and services among themselves for a variety of reasons; nearly all now buy some foodstuffs from others and some buy a substantial part of their basic food requirements; income elasticities of demand for food tend to be high and to rise with income; and effective elasticity of demand at retail is much enhanced by the shift from consumption of home-produced food to consumption of purchased food. The effective demand faced by farmers is of course influenced by rising incomes and increased reliance on the market for food supplies. It is also affected directly and indirectly by the performance of the marketing system: directly as the spread between farm prices and retail prices is altered and indirectly as the marketing system's ability to provide reliable supplies at reasonable prices makes consumers more or less willing to rely on it for their basic requirements. Collinson (1972:29) calls the exchange facilities of the market "perhaps the most effective catalyst for system evolution." Even before markets existed, however, variation in staple crop yields among households made necessary "reciprocal responsibilities between the household and the group to alleviate individual food shortages" and to balance supplies within the group (Collinson:27). Household self-sufficiency probably never existed. Community self-sufficiency may have, but at considerable cost.

The efficiency and capacity of existing agricultural marketing systems are of critical importance for the development of African agriculture, as they determine both the extent to which farmers can expect to be able to sell all they produce at a reasonable price and the ability of consumers to count on finding supplies when they need them, also at reasonable prices. As Heyer points out, "The main reason why subsistence farmers [in Kenya] do not use the market is nearly always that the market is too variable with respect to prices and sometimes also with respect to supplies to be relied on" (Heyer 1965:3).[29]

Anthony and Uchendu believe that Geita District, Tanzania, could become a major food-producing area, but that this would require "an efficient marketing system, [and] less emphasis on regional self-sufficiency." Markets seriously limit the expansion of crop production, and "local markets for foodstuffs can be easily glutted with only a small increase in production. . . . Until a wider market comes into existence and the distribution system is better developed there will be recurring local food shortages and wide fluctuation in food crop prices" (Anthony and Uchendu 1974:72–73).

A good marketing system must have the capacity to provide regular outlets for the products of farm and factory, regular supplies of foodstuffs and a variety of other consumer goods in both town and country markets, and timely supplies of the productive inputs needed by farmers and of the raw materials needed by processors and manufacturers.

Of the five food-marketing systems studied by Jones and his associates, those of southern Nigeria are the most highly developed.[30] Staple food marketing in southeastern Ivory Coast, southern Ghana, Togo, and Benin is probably comparable. Before Zaïre became independent, the supply hinterland for Kinshasa was also served by a similar system. The least developed of the food-marketing systems examined was that in Sierra Leone, where only about 6 percent of the domestic rice crop was marketed, although rice is the dominant staple. Imports of rice by Sierra Leone in 1966 were more than twice the amount of the local crop that was sold. The principal markets for all domestically produced foodstuffs are a few large towns and the diamond and iron mines, with very few purchases by rural people.

Food marketing in Kenya presents a different picture. There the enthusiasm for statutory marketing boards resulted in the establishment of no less than twenty-seven of them by 1960. The East Africa Royal Commission of 1953-55 said of these boards that they imposed "a degree of inflexi-

[29]She goes on to assert that official control of staple food marketing, as in Kenya, Tanzania, and Rhodesia, tends to reinforce the variability (Heyer 1965:3).

[30]The staple food supply hinterlands of Freetown, Ibadan, Enugu, Kano, and Nairobi. Similar studies have been made of Ethiopian markets by A. R. Thodey (1969) and by Winifred Manig (1973), of Ghana markets by V. K. Nyanteng (1969), and of Nigerian markets by J. O. C. Onyemelukwe (1970) and S. O. Onakomaiya (1970). Data for Onyemelukwe's dissertation were collected before the outbreak of the Nigerian Civil War, at the same time that Anita Whitney (1968) was studying eastern Nigerian markets.

bility which inhibits the economic advancement which is desired'' (East Africa Royal Commission 1955). These relics of colonialism were retained with little change by the government of an independent Kenya, as were the export marketing boards of other former British dependencies, including Nigeria. Kenya differed from Nigeria in the number of its boards and in their participation in internal trade in foodstuffs, not a function of the boards in colonial West Africa.[31]

In an article published in 1970, Jones concluded that ''in terms of the tasks that marketing systems are asked to perform, the African ones that we studied are not performing badly.'' Sierra Leone suffered from an inadequate transportation system and Kenya merchants were badly crippled by government intervention, but still managed to market perhaps four times as much maize in a year as the marketing board. The marketing systems were adjusting to the introduction of new crops and growing urban demand (Jones 1970:192–93).

In general, the studies showed many of the allegations that have been made about African marketing systems to be ungrounded. The average rise in prices between harvests seems roughly equal to the cost of storing commodities for this period.[32] The market chain between producer and consumer of most commodities is quite short, and net returns to intermediaries are modest. Producers and consumers alike usually have numerous alternatives when selling or buying, and there is little evidence that traders can or do exploit farmers. In many instances where statutory boards have legal monopolies of grain marketing, buyers and sellers prefer to deal with private merchants who illicitly offer prices that are more attractive than the official ones. In major centers of population, supplies appeared to be adequate throughout the year and the spread between farm price and consumer price moderate. Farmers' access to markets for their produce varied greatly with their location, the extensiveness of road networks, and the extent to which merchants could expect to find supplies in a particular area. There was little evidence that ethnicity or the extended family constituted barriers to smooth functioning of the marketing system, and on the

[31]For discussion of marketing boards in West Africa, primarily Ghana and Nigeria, see Peter Bauer (1954); for British East Africa, see East Africa Royal Commission (1955); for Sierra Leone, see R. G. Saylor (1967). Detailed official investigation of the Kenya Maize Board is reported in the Kenya, Maize Board Commission of Inquiry (1966). Peter Stutley provided a succinct summary of the difficulties and deficiencies of government intervention in agricultural marketing in the paper he presented at the Second International Seminar on Change in Agriculture at the University of Reading (Stutley 1976).

[32]Of course, the seasonal rise in any particular year may be much greater or less than this.

whole, farmers displayed a sophistication in economic exchange comparable to that shown by merchants.

Some deficiencies in the system do cause concern for future development. Even in the highly developed food-marketing economies of western
Africa, arbitrage among spatially separated markets was only fair, and
where road networks were poor, as they were in Sierra Leone, spatial
arbitrage was poor also. Impaired spatial arbitrage in Nigeria is associated
with problems of market access which could be overcome by development
of organized commodity markets in major supplying areas and urban ter-
minals, building on the existing brokerage system. Spatial arbitrage could
also be much improved if more concerns with national capacity were
engaged in the internal food trade. Some evidence suggests that prior to the
establishment of marketing boards and licensed buying agents for export
crops, the private traders in these commodities may also have played a role
in the internal food trade. This certainly is not unreasonable. Over most of
tropical Africa it has been the cash received from export crops, either
directly or as wages paid by plantations, that has been the major source of
market demand for all consumer goods, including the products of agriculture. The same mechanisms that made possible the evacuation of export
commodities were also used to provide things farmers wanted to buy, and
could have been used for the internal food trade.

## Export Marketing and Internal Marketing

When ocean transport developed between Western Europe and the Atlantic Coast of Africa, long-distance trade took on a new dimension. The
great export trade in the products of agriculture that dominated the economic history of Africa during the last seven decades did not begin until
near the end of the nineteenth century, and then it consisted largely of
products gathered from the forest—palm oil and rubber—that were later to
be grown on small farms and plantations. Collection and evacuation of
these new products required trails and roads, security, supplies, and ac-
commodations for merchants and their caravans, and carriers. They also
required that there be a sufficient company of merchants to seek out,
purchase, and transport the desired commodities—sometimes even persuade farmers to produce them—at a cost low enough so that prices paid to
farmers would induce them to supply the commodity and prices of the
product when delivered to the port would make it competitive in world
markets. As the number of commodities and their volume increased, the
network connecting producing areas to the ports became longer and more

complex. The first commodities came mainly from the forested areas—palm kernels and oil, rubber, cocoa, coffee, bananas, and kola. Peanuts and cotton also became important later.

But roads, transport, and marketing organizations—facilities set up to move one product—can also be used to move others, as was demonstrated when the "tea roads" were built in Kisii District, Kenya (Uchendu and Anthony 1975b:46). It should be no surprise to find a considerable correlation between the development of an export crop industry and that of an internal marketing system for staple foodstuffs, except when special circumstances (including marketing board monopolies) made it difficult to use the same routes and intermediaries for both or because the increased purchasing power generated by the new exports was not widely disseminated through the system.[33] (This relationship shows up in the studies by Jones and his colleagues.)

By their very magnitude, the great export trades assured access to markets and economical equilibration of differences in price over time and space. A similar result could probably be achieved in many parts of tropical Africa by greater reliance on and fostering of the existing brokerage system. Brokers are particularly common in Muslim western Africa, but are said once to have been common across the continent and down into the Congo. A system of resident commission agents or brokers can do a great deal toward overcoming deficiencies in knowledge and variation in local custom. The broker increases access to product markets and at the same time improves the bargaining position of small buyers and sellers. Thus the refinement and expansion of brokerage systems, together with better use of the large-volume trade in export commodities and in some products traded interregionally (e.g., maize in Kenya), might be the most effective way of increasing the capacity of domestic markets for agricultural products and of improving the efficiency with which they distribute commodities. The brokerage system could provide the informed market specialists, and the large-volume trades could provide the volume of business necessary to support the range of activities that should be carried out. Present actions by governments seem all too often to be in the opposite direction.

[33]Hunter's criticism of crop development authorities is particularly apt for most marketing boards (Hunter 1973:239): "Most of these organizations deal with a single crop or product, usually of high value (to cover overheads of management) and often, though not always, an export crop. In consequence, such corporations tend to be uninterested in the other crops which most farmers grow. The concept of farm management, as a means of maximizing total income by an optimum use of land, labor and other resources, is lost to view. A whole set of powerful, vertically organized, single-commodity organizations is very hard for either a ministry of agriculture or a farmer to deal with."

## Conclusion

It is possible to paint a very gloomy picture of domestic market prospects for tropical African farmers using the FAO figures on proportion of the population in agriculture (Table 2.1), assuming that the rural population is largely self-sufficient in foodstuffs, and estimating income elasticity of demand for local food at about 0.50 (Table 3.16). The reality is less discouraging. Statistics of agricultural population conceal a great diversity of occupations, many rural people already buy significant quantities of locally produced foodstuffs, and lively local trades have developed to meet their needs. The analysis by Stevens suggests rather convincingly that consumer demand for food may in fact increase much more rapidly than income.

All this argues that there exists a potential effective demand sufficient to support the specialization in production that will be necessary if output is to grow. Economic exchange is deeply imbedded in most African societies, and African merchants have persevered in its conduct despite the indifference and obstructionism of colonial governments. Rural populations are not an undifferentiated homogeneous mass—an amorphous peasantry in Polly Hill's words (1968)—but are made up of individuals with widely varying abilities and ambitions who are united by economic and social relationships. The rural population is deeply involved in the exchange economy through the things it buys and sells and through the income it receives from the migratory, part-time, and full-time wage employment of its members. In many parts of tropical Africa, farm families already command resources sufficient to finance profitable investments in farming.

None of this should obscure the fact, however, that only a small part of the potential domestic market demand for foodstuffs has been realized. The rate at which potential buyers can be brought into the market, especially for staple foods, will depend heavily on the efficiency of the internal marketing system for both domestic and export crops and on its ability to distribute greatly increased volumes of product. Over the long term the rate of the growth of the nonfarming population will be the most important determinant of the domestic demand. It will also be influenced by the success farmers have in reducing production costs (increasing productivity) so that more consumers can be won to the market. The time when local demand for farm products will be saturated and resources employed in agriculture reach a ceiling is very far away.

# 4. Farming Systems and Their Evolution

Because of variation in climate, soils, and topography, agricultural production in tropical Africa is carried on in a great number of microenvironments. Cultural and historical factors, often associated with particular ethnic groups, account for additional variations in farming practice and in the sociocultural environment. Nevertheless, despite wide variation in practices, African farming systems share common characteristics and problems.

## Traditional Systems: General Characteristics

A large fraction of the economically active population is engaged in agriculture, and small-scale family units account for the great bulk of farm output. Large European-owned farms or plantations have been important only in Rhodesia and to a lesser extent in Liberia, Zaïre, the Ivory Coast, Kenya, and Zambia. Farms of seven to ten acres commonly support families of five to seven people.

Farming is characterized by lack of capital and little use of purchased inputs. Consideration for subsistence needs is usually primary, and the low level of production of cash crops restricts the accumulation of capital and the use of inputs produced off the farm.[1]

The basic capital of the small farmer consists of a few huts and granaries, constructed by the farm family, a small range of implements for hand cultivation (axes, cutlasses, hoes, and knives), and equipment for food preparation. As enterprises are intensified, farmers may accumulate cash for the hire of labor and the purchase of seeds, fertilizers, insecticides, and spraying equipment. In savanna areas they may acquire oxen and ox-drawn equipment.

[1] Bede Okigbo says (1976:163) that "these systems... are very precisely adapted to enable farm families to get as much as they can of what they want, year in and year out, from their environment, using the resources they are able to devote to the agricultural sectors of their life systems. It is not surprising that technology based on studies of methods of production for single crops in isolation... have had so little effect.

Traditionally, land was not capital. Ownership of land was vested in the community or its leaders. Individual ownership was not important when land was plentiful. The basis of land rights was usually residence, and crops were recognized as individual property. Rights to fallow land were also often respected, but when land was deemed abandoned it reverted to the community. With increasing farm population, leading to land shortage, and the growth of cash crop economies, land has become increasingly valuable. Land tenure arrangements have changed, and continue to change, in response to these new influences. Where previously there was no market for land, it is now frequently bought and sold, though in many communities it is considered shameful for a man to sell family land.

Cattle are important only in the drier and tsetse-free savanna areas. But even in these areas they are used to only a limited extent as work animals and are not integrated into farming systems. Little use is made of animal manure, even where large numbers of animals are maintained. Farmers are deterred from its greater use by the labor requirements of manure conservation, its transport to the fields, and of spreading. Where manure is used, it is usually dry and leached by exposure to sun and rain, and is of limited value as fertilizer. The so-called compound farms of West Africa are a special case, where around homesteads or close to villages the land is cropped every year, and fertility is maintained by applying a mixture of animal manure and household refuse to the land. Elsewhere, bush and grass fallows are interspersed with periods of cropping to maintain crop yields at an acceptable level and to control weed growth.

Productivity per man and per acre is low in farming systems in which reliance is put upon natural fallows to restore fertility and in which there is little or no use of fertilizer and pest-control measures. Farmers recognize, and as far as possible use, the most suitable soil type for each crop or crop association, and each environment has its recognized cropping sequence. Food crops are usually sown as mixtures of species and varieties. Crop mixtures spread farm work requirements and provide some insurance against bad weather or pest attack, from which all crops and varieties do not suffer equally. Legumes, for example cowpeas, are frequently undersown and probably contribute to the supply of soil nitrogen.

The principal inputs are land and labor, and much of the labor is provided by the farm family, although in certain parts of Africa—particularly West Africa—hired labor is also extremely important. Tasks are often sex-linked. Some jobs are traditionally carried out by men, others by the women. Usually, men are responsible for clearing bush, the removal of tree stumps, and similar heavy duties, as well as for those activities linked to the man's role as taxpayer and producer of cash crops and of any other crops that farmers may be legally compelled to grow. Women are responsible for the food crops, fetching water and firewood, and for food preparation.[2] Joint tasks include land preparation and weeding. Practices vary from place to place; in Muslim communities, women do little farm work. Children help about the farm and are made responsible for bird scaring and herding animals.

Additional labor, made necessary by the expansion of farm operations or because of illness or other domestic problems, may be hired or obtained from one of the traditional sources of exchange labor. Traditional work groups are becoming less important than they used to be and, in most parts of Africa, are gradually disappearing. Work groups are organized by and serve neighbors, and provide labor on a rotational-aid basis or in exchange for hospitality. Societies formed for a purpose other than labor supply—for example, young men's dancing societies among the Sukuma—also provide labor for specific tasks. Unfortunately, although work groups accomplish a great amount of work in a short period, it is rarely of a high standard and the seasonality of farm work makes it difficult to obtain assistance at the time it is most needed.

Seasonal labor shortages are characteristic of systems heavily dependent on labor and with very different labor needs at different times. It is difficult to ensure that farm operations are carried out at the best of times. Land preparation, whether by hand implements or ox plow, is not easy to accomplish in the dry season when soils are hard and work animals are often in poor condition, and this operation tends to be delayed until the first rains. This delays the sowing of annual crops, the timing of which is particularly critical. Weeding and harvest times are also periods of peak requirement. In contrast, there are long periods when little field work is necessary.

Emphasis is placed on the production of the family's food needs. A marked increase in output and in the marketable surplus of subsistence crops can easily saturate the narrow local markets. As a result, there tend to be wide fluctuations in the prices of food crops. Accordingly, farmers are reluctant to rely on markets for the supply of their food needs, and give priority in times of labor scarcity to the cultivation of their food crops at the expense of work on the cash crops.

Where capital is scarce and implements are simple, differences in the cropped area of the farm reflect the land and labor resources available to the farm family. The efficiency of farm operations depends upon manage-

[2] Among the Yoruba and many other people of western Africa, cultivation is almost exclusively the responsibility of the men (cf. Forde 1950:153).

ment, which is extremely variable in a situation where the bulk of the population is engaged in farming. Individual action and change is limited by the social environment of family and community; for example, community pressure may exercise a veto on enclosure and inhibit attempts to improve livestock. Traditional practices cannot be readily ignored, and where a farmer may be convinced that it is to his advantage to adopt a new method of doing a farm operation, he still has to convince those who carry out the work. This may be his wife or a work group, who are accustomed to following established methods. In some communities having patriarchal systems of authority, such as that of the Kusasi in northern Ghana, all decision-making on the farm is vested in the senior and oldest members of the compound family, the persons least likely to be receptive to new ideas.

## Traditional Systems: Geographical Variation

The three main geographical zones of tropical Africa are forest, savanna, and highlands. Many of the special environmental problems facing small-scale farmers in tropical climates are common to all three zones.[3]

Rainfall and the availability of water for crops and livestock are of special importance to farming in tropical regions. In temperate climates, crop-growing seasons are determined by temperature and the occurrence of killing frosts at the beginning and end of the growing seasons. Except at high altitudes, temperatures in tropical Africa are usually adequate for crop growth throughout the year. The amount and distribution of rainfall is the major factor determining the season, the type of crop grown, and the geographical variation in the farming systems.

A. H. Bunting notes that the seasonal pattern of availability of soil water in tropical regions with a pronounced dry season is the opposite of that in temperate regions. In temperate climates, rainfall exceeds evaporation during the winter and the soil profile becomes charged with water. In the crop-growing season, evaporation commonly exceeds rainfall, but crops can use the reserves of soil water accumulated during the winter months. There is little leaching of nutrients beyond the crop-root zone during this period. In contrast, in tropical savanna areas, available soil water within crop-root range is usually exhausted by the end of the growing season. The profile is left with little or no reserves of water at the beginning of the

---

[3]General treatises like those by Webster and Wilson (1966), Angladette and Deschamps (1974), Wrigley (1969), and Ruthenberg (1976) are therefore relevant to understanding the farming systems found in the countries of tropical Africa. A collection of papers edited by Leakey and Wills (1977) is principally useful for West African agriculture.

following rains, and young crops are vulnerable to the short dry spells which frequently occur early in the growing season. Rainfall exceeds evaporation for long periods during the rains, and the profile is continuously leached (Bunting 1970).

During intense tropical storms, rainfall exceeds infiltration and much of the rain is lost as surface runoff. Runoff increases as the impact of fast-falling raindrops seals pores in the surface layers of the soil. Arable land is then without cover and grassland bare and overgrazed, with a hard, trampled surface. Flooding of low-lying places frequently occurs at the end of a period of drought, and erosion can be serious even on moderate slopes. Soil and water conservation measures have rarely proved popular with small-scale farmers. They are frequently troublesome to adopt and the advantages of doing so may not be immediately obvious. Colonial records abound with accounts of abortive campaigns to promote soil conservation, often involving compulsory measures at field and village levels.

Soil fertility is variable. In some areas soils have high intrinsic fertility, as, for example, the upper valley soils of Zambia. But extensive areas of tropical Africa are covered by highly weathered soils of low fertility. Available nitrogen and phosphate are often deficient and under natural fallow systems, with little or no fertilizer dressing, crop productivity is low.

The high temperatures of the tropics are favorable for plant growth but cause rapid growth of weeds as well as crops. The grasses *Digitaria scalarum, Imperata cylindrica, Cynodon dactylon,* and nut grass (*Cyperus* spp.) are particularly pernicious weeds of cultivation in tropical Africa. Sorghum production may often be seriously reduced by witchweed (*Striga,* spp.).[4] Inability to control weeds adequately by current methods is a major factor limiting crop production in Africa.

In the tropical environment, pests and diseases quickly multiply to epidemic proportions and cause severe crop loss. The spread of crop-sowing dates in the traditional farming systems gives rise to a sequence of plants of different ages and may encourage the buildup of pests in early sowings to the detriment of later sown plants.

The presence of large numbers of host plants provides major sources of crop infestations. For example, the importance of American bollworm, strainers, and lygus as pests of cotton is often related to the incidence of other hosts in the natural vegetation that are so numerous as to make eradication impracticable.

[4]*Striga* spp. are parasitic herbs able to attach themselves to the roots of their host on which they largely rely for food.

The complex relationship between climate, host, pest, and predator results in situations where a pest may cause major crop loss in one area and yet be relatively unimportant in another. Thus, *Heliothis armigera* is not of great importance in equatorial Africa, where equable conditions appear to result in a natural balance between the pest and its natural enemies. In contrast, *H. armigera* is a major pest of cotton and other crops in savanna areas, where it is better adapted to survive a lengthy dry season than its natural enemies.

Traditional farming systems were developed in response to the economic, environmental, and social influences noted above. Where there is adequate cultivable land, natural fallows are used to renew soil fertility after cropping. Plant nutrients are cycled between soil and vegetation; nutrients leached below the root zone of annual crops during the arable period are taken up by the deep roots of the vegetation during the fallow period and returned to the topsoil. Losses of nutrients to the soil-vegetation system occur by removal of crops, by leaching out of the root zone of the vegetation, and by erosion. The action of denitrifying organisms removes nitrogen from the soil, and volatilization and a loss of organic nitrogen and sulfur occurs when fallows are burned in preparation for cropping or, in the savanna areas, to induce the growth of fresh grazing. Nutrients are gained in rain and dust, and atmospheric nitrogen is fixed by microorganisms in the soil (Nye and Greenland, 1960).

Sowing mixtures of crops minimizes the period that land cleared for crops is left exposed to storms, and provides an insurance against the very serious crop loss to pests which a farmer might incur if he relied upon a single crop variety. Weeds are a major cause of low crop yields; they build up during cropping but are suppressed by regenerating vegetation during the period the land is under fallow. They are further checked by the burning of crop trash after harvest. Traditional systems are not highly productive, but they incorporate well-tried methods of cultivation and provide farmers with a means of ensuring the subsistence needs of their families.

## The Forest Zone

The closed forests of the lowland equatorial regions cover about one million of the five million square miles of tropical Africa south of the Sahara. P. H. Nye and D. J. Greenland identify closed forest with areas having an annual rainfall of fifty inches or more and a dry season of not more than four months. Moist semideciduous forest usually gives way to moist evergreen forest in areas receiving more than seventy-five inches of rainfall per annun (Nye and Greenland 1960). High rainfall, in some areas exceeding 200 inches per annum, is the major factor determining agricultural production and practices in the zone.

A pronounced dry season of about four months occurs on the west coast of Africa, but more typically the forest zone receives less than two inches of rainfall during only one or two months in the year. Climate is characterized by monotonous heat and humidity and abundant, even excessive, rainfall. Mean daily temperature in the lowlands is high, about 80°F., but extreme temperatures rarely exceed ± 10°F. of the mean.

Soils are predominantly deeply weathered and intensely leached latosols. Nutrient cations are largely held by the organic matter contained in the surface layers of the soil. Despite the low fertility of the latosols, they are able to maintain a cover of semideciduous forest by the circulation of a small capital reserve of nutrients in a closed soil-vegetation cycle. In areas supporting semideciduous forest, leaching is less than in the very wet areas and soils contain a higher proportion of organic matter and nutrients, particularly nitrogen, calcium, and magnesium, than do the less fertile soils of the evergreen forest areas.

Land is prepared for crops by cutting down and burning the vegetative cover and cultivating with a hoe. Clearing exposes soils to erosion until crops are sufficiently established to provide ground cover. Intense leaching and the removal of nutrients by crops lead to rapid loss of soil fertility during the arable phase of the cropping system. Yields of successive crops decline rapidly, and land has to be rested for long periods to restore fertility; cultivation periods are customarily two to six years and fallow periods, six to twenty years.

Traditional practices of mixed cropping ensure early and continued cover of the soil. A typical crop sequence begins with the sowing of a cereal, usually maize or rice or a mixture of the two, interplanted with minor crops, including chillies and vegetables. This mixture is later interplanted with bananas and semiperennial roots like cassava, yams, or cocoyams, which provide a vegetative cover after the cereals are removed. They are harvested in the following and subsequent years until regenerating bush fallow takes over. In the cocoa-growing areas of Ghana, it is common practice to plant the major food crops (plantain and cocoyams, with maize interplanted in the first year) on the same land as new cocoa. The food crops provide a return from the land until the cocoa comes into bearing. The calorie yields per acre and per man-hour of the root crops are high, but their protein content is low. Consequently, diets in the rain forest zone are likely to rely heavily on supplementary sources of vegetable protein like green leaves, pulses, and peanuts. Few animals are kept and livestock are confined to a few goats, sheep, and chickens.

Tree crops are ideally suited to the high rainfall conditions of the forest zones, and are grown extensively. Their leaf canopy breaks up the rain-drops of the intense tropical storms and preserves soil structure. Their extensive root systems hold the soil in place and prevent constant erosion, as well as enabling the trees to withstand considerable periods of drought. The main tree crops of tropical Africa are cocoa, coffee, tea, oil palms, coconuts, kola nuts, and rubber. Oil palms and rubber thrive best in the wetter areas where rainfall is evenly distributed throughout the year.

## The Savanna Zone

The savanna zone extends from the tropical rain forest to the semiarid areas adjoining desert, and covers areas with mean annual rainfall from sixteen to eighty inches and with a well-defined dry season. The zone includes a range of natural vegetation types which reflect both climate and the influence of man.

A mosaic of savanna vegetation with forest remnants borders the forest zone in areas having an annual rainfall of more than forty-five inches, falling in one or two rainy seasons, and a dry season of three to five months. Tsetse fly limits cattle rearing in the region and, as in the forest, a few goats, sheep, and fowl are the only animals kept. The dominant crops are cereals (maize, finger millet, and sorghum) and cassava. Yams are a major food crop in West Africa and are often the main crop grown in the year land is opened for cultivation. Bananas are important in the elephant grass areas of Uganda and are also a major crop in Rwanda, Burundi, in forest and highland areas in Zaïre, and in some localities in West Africa. Secondary food crops include peanuts, sesame, bambarra nuts, beans, and a range of vegetable crops. The main cash crops are Robusta coffee and cotton. Cocoa is grown in wetter areas adjoining the forest zone.

In areas having a rainfall of twenty-five to fifty-five inches, the major vegetation formations are various types of tall-grass savanna, consisting of mixed collections of trees with high fire tolerance and a tall-grass ground cover. Thorny species increase as conditions become drier. Centuries of burning, clearing for cultivation, and grazing have modified the vegetation of large areas and resulted in parklike landscapes with widely spaced trees, composed of species which are valued for their economic worth.[5] The parkland of the West African Sudan savanna zone and the cultivation steppe of Sukumaland are typical of these areas.

The main crops are sorghum, maize, peanuts, bambarra nuts, beans,

[5]In West Africa typically, *Butyrospermum paradoxom*, *Parkia filicoidea*, and *Tamarindus indica*.

cowpeas, sweet potatoes, and cassava. Maize, cotton, peanuts, and tobacco are important cash crops; coffee is grown in wetter areas. A common
crop sequence is cotton (in the first year) followed by a food crop mixture
for two or more years. Typical crop mixtures for the Sukumaland area of
Tanzania are: (1) cassava, maize, peanuts, and bambarra nuts; (2) cassava,
maize, sweet potatoes, beans, and cowpeas; (3) sorghum, millet, and sweet
potatoes; and (4) sorghum, millet, and legumes. Cultivated land is commonly allowed to revert to bush under cassava or pigeon pea planted
toward the end of the crop sequence.

Where the absence of tsetse fly permits, large herds of cattle may be
maintained by settled farmers, as with the Iteso of Uganda, the Plateau
Tonga of Zambia, and the Sukuma of Tanzania. In these areas, ox cultivation has been adopted to varying degrees, within the traditional systems of
shifting cultivation. The tall-grass savannas are also suitable for mechanical cultivation, and in southern and central Africa the more fertile areas are
successfully farmed by medium- and large-scale farmers.

Miombo woodland and savanna, dominated by *Brachystegia, Julber-
mardia,* and *Isoberlinia* spp., with ground cover of herbs and grasses
subject to annual burning, is widespread south of the equator in areas with
twenty-five to thirty-five inches of rainfall and a dry season of six months
or more. Because of poor soils, unreliable rainfall, and poor water
supplies, the human carrying capacity of miombo country is low with
unaided subsistence farming (Allan 1965). In some areas, cultivators practice the *citemene* system of farming, in which fertility is created on land to
be cropped by the burning of brushwood collected from a large area of
woodland. Crops are similar to those in the more fertile tall-grass savanna
areas, but yields are usually lower. Ample but largely low-quality grazing
exists for cattle in the rainy season, but tsetse fly is widespread over vast
areas.

Short-grass savannas are typical of areas where the average annual rainfall is fifteen to thirty inches and concentrated in one rainy season with a
dry period of seven or eight months. *Acacia* spp. are frequently dominant
and have a ground cover of short grass. Herding of cattle, sheep, and goats
is an important way of life. Bulrush millet, sorghum, groundnuts, and
beans are the main crops, but because rainfall is low and variable crop
yields are usually poor. Groundnuts may be marketed in the wetter areas
but are largely grown for subsistence needs.

Cereals are the staple foods of the savanna areas, choice of cereal depending on availability of water, local food preference, the possibility of
bird damage, and the labor needs of the crop. Bulrush millet is the staple

food of the areas with ten inches of rain or less, but sorghum is usually the preferred staple of the dry savannas, and its cultivation begins at about the fifteen-inch isohyet. Finger millet, which requires more reliable rainfall than sorghum, is important in some of the more humid savanna areas, largely because of its excellent storage properties.

Maize is frequently a preferred cereal. The crop commends itself to farmers, partly because of consumer preference but also because it can give a high return to labor, is easy to harvest, and can be harvested over a long period. Its husk gives protection against rain as well as birds. Maize is sown where rainfall permits, but is less drought tolerant than sorghum, for which it is increasingly substituted in the savanna areas.

Because of its water needs, rice is largely confined to riverine areas or to upland areas with well-distributed rainfall. Upland rice, which accounts for most of the rice grown in tropical Africa, is almost entirely grown in rain forest areas.[6]

Where cattle are important they have a multiple role. They provide food, may be trained as work animals, and, traditionally, play an important part in the social life of cattle-owning peoples. Although the social significance of cattle is less important in a rapidly changing Africa, large numbers of cattle still give prestige to the owners. They are a means of storing wealth and are important items in marriage exchanges. In some communities the complexity of ownership rights of animals, acquired by marriage or exchange or inheritance, makes it difficult to sell beasts. In West Africa, however, there is a large movement of cattle from the grasslands of the savannas to the cities near the coast. Herds are allowed to increase naturally and are not culled. Until recently, ownership of a large herd made the owner less susceptible to impoverishment by a long period of drought or an outbreak of disease.

Animals may be grazed on other farmers' crop fallows or crop stubbles, provided no damage is done to the neighbor's property, and for this reason the introduction of farm enclosure is usually vigorously resisted. Cattle owners largely rely for forage upon unimproved natural grazing. Where land is available, grazing rights are usually communal and herds are pastured together, sometimes under the care of professional graziers such as the Fulani of West Africa and the Nkole of Uganda.

Herds are moved seasonally to areas providing the best grazing, but the stock-carrying capacity of unimproved savannas is low and twenty to

---

[6]Paul Pélissier provides a particularly rich description of the ancient rice culture of the Casamance area of Senegal, based originally on *Oryza glaberrima* (Pélissier 1966:621–886).

twenty-five acres or more are required to maintain one animal (Webster and Wilson 1966). Water supplies are usually poor and localized overstocking commonly occurs around watering points. The practice of burning grazing land during the dry season in order to induce new vegetation growth results in uncontrolled fires which lead to continued deterioration of the vegetation over large areas and expose land to soil erosion.[7] Animals are subject to biting flies, ticks, a wide range of internal parasites, and diseases; some of these, such as rinderpest and East Coast fever, are specific to the tropics.

Indigenous cattle are hardy and adapted to harsh conditions. In the rare instances where selection of animals has been practiced under traditional systems of cattle rearing, it has been for noneconomic characteristics. Animals are much less responsive to good management, notably improved nutrition, than temperate breeds.

Natural fallow systems of farming are practiced throughout the savanna zone. In the savanna areas the dominantly grass fallows are subject to annual burning with accompanying loss of nutrients, slowing down of humus accumulation, and exposure of the soil to erosion at the beginning of the rains. Burning is most fierce and destructive to tree species when it occurs late in the dry season; then only fire-tolerant species survive. Where fallows are burned late and periods of regeneration are short, the grass *Imperata cylindrica* becomes dominant at the expense of woody species and is difficult to eradicate during the cultivation period.

The ratio of fallow to cultivated area is determined by population pressure and soil fertility. Latosols are the most common soil type and large areas have a lateritic pan above the rotting rock. Nitrogen and sulfur are commonly in short supply and limiting to crop yields. Phosphate is limiting on large areas, especially when these have been subject to long periods of cultivation, but the store of soil potash is usually adequate. There is a great range of fertility among soil types. On some poor soils in Zambia, notably soils of the miombo woodlands, two or three years of cultivation have to be followed by a fallow of twenty to thirty years in order to restore soil fertility. In contrast, on the fertile upper-valley soils of Zambia, after a long cultivation period of four to ten years, fertility can be restored by a fallow period of the same length or shorter. In some East African systems, such as that of the Buganda of southern Uganda, a cultivation period of

[7]Controlled burning at the end of the dry season, especially if carried out after the first showers, is a satisfactory method of controlling bush growth, but it must be accompanied by good management (Webster and Wilson 1966).

three years is commonly followed by a rest period of the same length (Allan 1965).

Compound farming is practiced over large areas of West African savanna in farm enterprises which also include natural fallows. This system of maintaining permanent cultivation on land by the application of household refuse and manure possibly arose in response to the need for large numbers of people to live close together for ease of defense.

## The Highland Zone

Extensive and agriculturally important highland areas occur in East Africa and Ethiopia. Large areas are densely inhabited up to 8,000 feet. High concentrations of population are made possible by a favorable environment, adequate and reliable rainfall, plentiful supplies of surface water, and fertile soils, although in Ethiopia especially a long period of cropping has led to substantial loss of soil nutrients. The carrying capacity of the land has encouraged population growth, and early colonization of some areas was prompted by the greater security from warring neighbors offered by the mountainous terrain.

Big differences in rainfall and temperature are experienced within the zone. In the Kenya highlands, mean rainfall varies from about thirty to seventy inches per annum. Temperatures decrease by 3.3°F. for every 1,000-foot increase in altitude, and where it is too cold for crops, land is used only for grazing. Although only small annual variations in mean monthly air temperatures occur on the equator, they can prove critical for crop growth at higher altitudes and restrict the growing season. Topography is often difficult for farming and soil conservation works are necessary to prevent serious erosion. Slopes are frequently too steep for ox and tractor cultivation.

High-value cash crops have provided opportunities for profitable farming. Arabica coffee is typically grown at altitudes between 4,000 and about 6,500 feet. The recommended lower limit for pyrethrum growing in Kenya is 6,500 feet, and the best altitude for crop flowering is 8,000 feet or more. Tea does best in areas with acid soil and high rainfall and these conditions tend to be found between 5,000 and 7,400 feet in Kenya, although wetter conditions enable the crop to be grown satisfactorily at 4,000 feet in Uganda, Tanzania, and Malawi. The main food crops are maize, finger millet, sweet potatoes, and beans, all sown in customary crop mixtures. Because of low temperatures, maize matures late at high altitudes and is unsuitable for double cropping even where rainfall permits. Bananas, cassava, groundnuts, and cowpeas, all of which need a warm climate, are usually grown at altitudes below 5,000 feet. Temperate crops can be produced satisfactorily and potatoes are grown both as a subsistence and a cash crop between 5,000 and 9,000 feet. Wheat is an important crop in the East African highlands, and in the Ethiopian highlands wheat and barley are both important but are outranked by teff (*Eragrostis abyssinica*), a distinctive indigenous cereal. Temperate breeds of livestock thrive in the subtropical climate of the highlands, and have been successfully introduced to small mixed farms, especially in Kenya. Well-cared-for pedigreed animals are to be found on farms of only seven to ten acres, sometimes with planted pastures.

## Modifications to Traditional Systems of Farming

Modifications to traditional farming systems have been influenced by demographic factors, economic incentives, and techniques available to farmers. A number of distinct farming systems, all small-scale, can now be identified. These include traditional fallow systems, compound farming, mixed farming, cash tree farming, and other less widespread systems, for example specialized horticulture and floodland cultivation.[8]

Shifting cultivation systems were developed by the earliest farmers and were well suited to areas having low population density, plentiful land resources, and a low level of technology. Land is cleared by slash and burn methods, a few harvests taken, and the clearing then abandoned. At intervals, farmers shift their homes as well as their fields.[9] William Allan distinguishes between obligatory and voluntary shifts. On poor soils, requiring a long regeneration period after cultivation, successive clearings are progressively more distant from the homestead. The journey to and from the fields becomes increasingly time-absorbing, makes the transport of produce laborious, and compels the farmer to build a new homestead nearer to his fields. Shifts are made not only from agricultural necessity. Farmers may move because of death or sickness in the family, or to a place where hunting is better, or simply because they are bored with their present location. Where population is still sparse and land plentiful, as in areas of

[8]George Benneh (1973) provides a recent classification of farming systems, with West African examples.

[9]The term ''shifting cultivation'' is frequently used to embrace all farming systems in which fields are periodically moved to a new site. In this chapter the term is only applied when holdings are also moved.

the River Congo Basin and on the Kalahari Sands of central and south-central Africa, there may be no need for the farmer to return to land that has been previously cultivated by him (Allan 1965).

As population increases and the widespread planting of cash crops results in a shortage of land suitable for shifting cultivation, farmers are compelled to rotate their crop clearances within confined areas. In later and more developed systems, the farmer lives in a fixed homestead or village and rotates his crops and fallows in what has variously been called natural fallow, rotational bush fallow, and recurrent cultivation systems. The pressures imposed by growing populations and by settlement patterns that result in localized high concentrations of people lead to further intensification of the system. The use of household refuse and animal manure to maintain the productivity of annually cropped land near the compound has already been mentioned. Similarly, in the yam zone of West Africa, much effort is put into creating suitable growing conditions for the staple yam crop, which needs reasonable fertility and loose soil. The crop is planted on mounds or hills made by collecting surface soil from the surrounding area and adding organic matter. In parts of southern Tanzania and Zambia, the *citemene* system is used to obtain comparatively good yields from soils of low intrinsic fertility.

Various government schemes have aimed at the development of mixed farming systems, notably the mixed farming scheme in Nigeria, but none has resulted in full integration of crops and livestock. Government schemes have a very mixed record of success, but the consolidation and resettlement program in the Kenya highlands has made good progress and shows what can be achieved by the implementation of a carefully planned scheme that is readily acceptable to farmers. The program has resulted in improved farms which incorporate intensive milk and permanent cash crop enterprises, and has been associated with the adoption of new technologies by farmers. Elsewhere, economic incentives, which have played a major part in influencing change, have led to specialized annual and tree cropping and, with government help, development of seasonally flooded and swamp areas for rice and vegetable production.

## Population Pressure

Population growth in tropical Africa has resulted from the success of efforts to reduce mortality, as well as from the increase in food production that resulted from the spread of New World crops and the increase in efficiency and effectiveness of famine relief under the colonial regime. As noted earlier, over much of the continent the rate of population growth probably exceeds 2.5 percent per annum. In the tsetse-fly savanna areas, the ending of widespread cattle raiding and the introduction of prophylactic injections have resulted in a similar increase in herds, which in the last century had often been decimated by theft and disease. The provision of drilled wells[10] enabled greater concentrations of animals to be kept than before, increased the pressure on available land, and led to overgrazing. The increase in numbers of animals has probably not resulted in a correspondingly large increase in food supply.

The carrying capacity of land under traditional farming methods is increasingly exceeded, and established ratios of cropland to fallow are no longer maintained. From necessity, land has to be cropped for longer periods and at the expense of time that it is under fallow. Reduction of the fallow period is accompanied by declining soil fertility and lower crop yields. There is less vigorous regrowth of fallow, and, in the drier areas, forest is progressively replaced by savanna. The grass fallows that result are probably less efficient restorers of fertility than the bush fallows they replace.

About 0.5 to 1.0 acre of crops per person is needed to provide minimum subsistence needs in tropical Africa. In very fertile areas, where bananas are the staple food crop, less than twice the cropped area may need to be included in the crop-fallow rotation. But in forest areas, where a long resting period has to be provided, more than ten times the cropped area may be needed. Allan has adapted the concept of carrying capacity of land, which is frequently used in range management to provide a crude measure of the number of people that a given kind of land can support under given technology, based on the food calories that it will yield. Using this approach, he estimates that the highly fertile areas like those on the slopes of Mt. Elgon, Uganda, can carry a population of as many as 600 persons per square mile if planted to bananas. The best environments may be capable of supporting 800 to 1,000 persons per square mile. In contrast, in some miombo woodland areas of Zambia, only four to ten persons can be supported per square mile (Allan 1965:88–89).

The effect of increasing population density on farming systems in a moderately fertile savanna area has been studied in Teso District, Uganda. Teso has good soil and an average rainfall of fifty-five inches per annum, with a bimodal pattern of distribution. There are two crop seasons, but the bulk of sowings is carried out in the earlier and longer rainy season. The main crops are finger millet, sorghum, cotton, groundnuts, beans, sweet potatoes, and cassava. Strip cropping was introduced to the area in the

[10]Called "boreholes" in former British Africa.

1930s, and by 1941 was practiced over most of the area. Crop strips thirty-five yards wide are separated by three-yard-wide grass filter strips.

The Iteso were formerly pastoralists and still maintain large herds of cattle and flocks of goats and sheep. Before the beginning of the present century, they practiced a simple crop system. The main concern was culti-vation of subsistence crops by women. Land was abundant and individual ownership not important. By the 1930s, land was customarily cropped for two or three years, followed by two and a half to seven years of rest. In a small sample of Teso farms, 40 percent of the total farm area was cropped in 1937 (Wilson and Watson 1956). About 50 to 60 percent of the land is cultivable, and Allan estimated in 1965 that Teso District could support an average of 100 to 150 persons per square mile (Allan 1965:189). The district has an area of 4,300 square miles and in 1930 had a total population of about 283,000 people, a density of 66 persons per square mile. By 1965 the population had doubled; the total population was 550,000 and the mean population density 130 (Uganda 1965).

The distribution of the population in the district is uneven. The southern counties, once noted as having the greatest potential for development, support the greatest population densities. Ngora, Kumi, and Bukedea had population densities of 240, 177, and 166 persons per square mile respect-ively in 1965. Pressure of human and livestock numbers had become criti-cal in Ngora County. Some voluntary migration had taken place within and beyond the district to relieve pressure on the land, but by 1966 had virtually stopped. Migration is a traditional method of adjusting growing populations to available resources, but provides no satisfactory answer once land is in short supply.

Samples of progressive farmers interviewed by Uchendu and Anthony were found to have cropped 44 and 57 percent of their holdings in Serere and Soroti counties in southwestern Teso District and Ngora County in southeastern Teso District in 1966.[11] The corresponding sample of neighbors of the selected progressive farmers in Serere and Soroti counties cropped an average of 77 percent of their holdings (Table 4.1). In Ngora County, neighbors cropped 94 percent of their holdings. Continuous crop-ping had become the pattern in Ngora, and land could no longer be rested between crops. Grass filter strips were increasingly encroached upon, and some farmers cropped the whole of their cultivable land. In other parts of southern Teso, land was cropped for two or three years and fallowed for

[11]Progressive farmers are farmers who have been noted as receptive to new ideas and selected by the extension services to receive special attention and help in the hope that they will catalyze change among their neighbors. Progressive farmers usually have larger and more prosperous farms than their neighbors.

two. In northern Teso, it was still possible to fallow for three or four years.

Land hunger in Ngora had created a market for land, and nine of the twenty-four farmers interviewed in the county had rented land for cash payment. Farms were becoming fragmented. They were desperately small, and the area devoted to cash crops and the gross farm income correspondingly low. The net return per farm for the sample farmers in Ngora was only a third of that of the sample in Serere and Soroti counties (Uchendu and Anthony 1975a).

Overcrowding in a poor savanna area, such as Bawku District in northeast Ghana, can perpetuate a bare subsistence economy. Bawku, in intermediate Sudan-Guinea savanna, has a mean rainfall of about forty inches per annum in a single seven-month rainy season. At the 1960 census, the population was 174,000 persons, with a mean density of 146 per square mile. In some localities population density is as low as forty to fifty persons per square mile, but where the proportion of cultivable land is high the density may exceed 600 persons per square mile. Available data indicate that the population is growing very slowly. This is probably related to the

*Table 4.1.* Farm data, Teso District, Uganda, 1966*

|  | Southwestern Teso Serere/Soroti | Southeastern Teso Ngora |
|---|---|---|
| Number of farmers in sample | 21 | 24 |
| Persons per farm | 10.2 | 6.2 |
| Farm size (*acres*) | 18.1 | 6.5 |
| Largest farm (*acres*) | 50 | 14 |
| Smallest farm (*acres*) | 4 | 2.5 |
| Number of blocks per farm | 1.1 | 1.3 |
| Area under cotton (*acres*) | 4.8 | 1.7 |
| Area under all crops (*acres*) | 13.9 | 6.1 |
| Percent of holding cropped | 77 | 94 |
| Acres cropped per person | 1.36 | 0.98 |
| Cash inputs (*shillings*) | 218 | 55 |
| (of which labor) | (149) | (23) |
| Cash returns (*shillings*) |  |  |
| Cotton | 1,166 | 295 |
| Groundnuts |  | 8 |
| Cattle | 69 | 100 |
| Other crops | 20 | 13 |
| Gross returns | 1,255 | 416 |
| Net returns | 1,037 | 361 |

*Data from Victor C. Uchendu and Kenneth R. M. Anthony, *Agricultural Change in Teso District, Uganda* (Nairobi, 1975a). Those interviewed were neighbors of progressive farmers.

poor resources of the area, which have encouraged emigration from the district (Uchendu and Anthony 1969).

There are three types of farm units in Bawku: compound farms, intermediate farms, and bush farms (Lynn, 1937).[12] The compound farm surrounds the dwelling and receives most attention. Manure is applied to it each year, but it is in limited supply and not more than 25 percent of the area is manured in any one year. The intermediate farm is a little distance away, seldom manured, and often fragmented. Bush farms are usually small, sited at some distance from the homestead, and on the poorest soils. Little work is done on bush farms beyond sowing seed and hoeing the growing crop once. Where land is sufficient, short periods of cultivation are rotated with long periods of fallow. Small satellite plots are also cultivated by individual members of the family to provide a private cash income. The cropping pattern is relatively simple. Early varieties of millet dominate crop mixtures sown in the more fertile areas on the compound farm adjacent to the homestead. Late millet varieties and sorghum are sown in the unmanured part of the compound farm and in the intermediate and bush farms. Data for a sample of sixty farmers in Bawku District in 1967, summarized in Table 4.2, show the low farm income, which is derived mainly from the sale of surplus peanuts and rice.

Cultivable land in Bawku District is poor, scarce, and overpopulated. Labor is abundant and there is extensive seasonal unemployment. Crop yields are poor. Typical yields per acre of the main food crops at Manga Experimental Station in the period 1952–66 were: early and late millet, 450 and 350 pounds grain, respectively, and peanuts, 450 pounds shelled nuts. Yields on farmers' bush fields were probably lower. Very little land can be spared for resting. Bush farms are small and far from farm homesteads. In some areas, such as Manga and Pusiga subdistricts, bush farms are now practically nonexistent. The use of household refuse and manure maintains a slightly higher level of yield on compound farms, but the supply of animal manure is limited. Only small numbers of sheep, goats, and poultry are kept, and few households own cattle. Crop production barely exceeds the family's food needs. Any small surpluses of groundnuts and rice are sold, but gross income is small and barely covers the needs of taxation and the small purchases of consumer goods. Close to the maximum number of people are supported at near subsistence level, and the agricultural population can expand further only if crop yields are correspondingly increased.

[12]Over much of tropical Africa, the word *farm* refers to an individual field. One farmer may have numerous "farms."

Table 4.2. Farm data, Bawku District, Ghana, 1967*

| | |
|---|---|
| Number of farms sampled | 60 |
| Persons per farm | 9.4 |
| Farm size (acres) | |
|     Compound | 5.3 |
|     Intermediate | 3.5 |
|     Bush | 1.0 |
|     Satellite | 0.9 |
|         Total farm size | 10.7 |
| Area cropped (acres) | 10.4 |
| Area under groundnuts and rice (acres) | 1.5 |
| Area cropped per person (acres) | 1.1 |
| Gross farm income (N₵)[a] | 28 |
| Cash inputs (N₵) | 6 |
| Net farm income (N₵) | 22 |

*Data from V. C. Uchendu and K. R. M. Anthony, "Field Study of Agricultural Change: Bawku District, Ghana," Food Research Institute Study of Economic, Cultural, and Technical Determinants of Agricultural Change in Tropical Africa, Preliminary Report No. 7 (Stanford, 1969).
[a]One new cedi (N₵) was equal to approximately $1.00 U.S. in 1967.

Excluding satellite plots, there are an average of three fragments of land per holding. However, single fragments of bush and intermediate farms are not so small as to present a major constraint to the adoption of yield-increasing measures such as improved seed, fertilizer, and pesticide. Farmers have the technology to use manures, but supplies are small. Fertilizers could be substituted if farmers could pay for them by the sale of their produce. The key is an efficient marketing system and a demand for Kusasi farm produce.

Uneconomic fragmentation of farms is commonly associated with high population density. Fragmentation of holdings usually results from the way in which land is inherited, but topography and the ease with which land can be bought and sold often contribute, as in Kigezi, a mountainous district in southwest Uganda. The birthrate in Kigezi is high, and increasing population has led to increasing numbers of small scattered fields, while family farms become progressively smaller (Bayagagaire 1962). It has fertile soils and, depending on location, receives an average of between thirty-five and seventy inches of rainfall per annum. About half the area is cultivable. The district had a population of 534,000 people in 1962, an average density of about 300 persons per square mile. The most populous area supported more than 800 per square mile. Sorghum is the chief crop of the highlands. Beans, maize, cassava, and coffee are also important.

A system of inheritance which led to excessive subdivision of property and which originated when land was more abundant, and a land-use system which developed in response to a need to include markedly different ecological conditions in the cropping pattern, have resulted in seriously fragmented holdings. In 1963/64, the average size of a farm in Kigezi was 6.0 acres, of which 4.5 acres were under cultivation and supported a family of 6.2 persons. There was an average of 6.2 blocks of land per farm. Less than one-third of the farms consisted of one block only, one-half had four or more, and some farms were made up of more than forty, each block only a fraction of an acre. Men who inherited uneconomic holdings had to leave their home area or lease whatever land they could, usually also in small scattered parcels. Land can be rested for only short periods of a few months or not at all. There is increasing cultivation on steep hill slopes, and contour strips are frequently dug up to provide more arable land. Without consolidation of holdings, it will be difficult to improve land. Plots are too small for enclosure and the maintenance of livestock, and it becomes impracticable to use fertilizer, manure, and insecticides on fields scattered miles apart. High-value vegetables, which can be produced on small plots and respond to intensive care, are the only field crops which can provide a worthwhile return under these special conditions.

Without fragmentation, a densely populated but fertile mountain area can still be highly productive. In Kisii District, Kenya, the mean population density is 700 per square mile, and in some localities 1,000 per square mile. Farms are as small as those in Kigezi but because they have been settled rather recently, are little fragmented. A range of soil types is assured to farmers by the layout of holdings, which are oriented as narrow strips from the bottom to the top of the ridge. Kisii farmers grow high-value cash crops, maintain productive pedigreed cattle, and are highly receptive to new ideas and change.

## Adoption of New Crops

Many of the commonly grown food and cash crops of tropical Africa have been introduced from other continents and absorbed into local farming systems. Bananas were an early introduction from Asia. New World species, which were first introduced by traders in the sixteenth century and gradually became established as important crops in the continent, include maize, peanuts, cassava, sweet potatoes, and tobacco. Most of the New World crops had spread widely across tropical Africa before the establishment of European rule. Manioc, for example, was present throughout the Congo Basin when the first Europeans traversed it (Jones 1959:60–87). The introduction of cocoa to Ghana in 1879, via the offshore islands, led to West Africa becoming the major producer of world cocoa. Upland cotton varieties from the United States were brought to West and East Africa at the turn of the century, and cotton production now makes a substantial contribution to the economies of many African countries. Arabica and Robusta coffee, first cultivated as large-estate crops, were introduced to small farmers in this century. Tea has been made available to small farmers in the highland areas even more recently.

A new crop receives wide acceptance where farmers recognize clear advantages to growing it. This might be because it provides a satisfactory cash return for an acceptable amount of labor or because a new food crop satisfies consumer preference, stores exceptionally well, is not readily attacked by pests, or provides a large return of produce per acre or per man-day of effort. A government promotion program may introduce a crop to an area, but its widespread adoption will depend on farmers' assessments of the relative advantages and disadvantages of growing it. The spread of cassava, and to a lesser extent of sweet potatoes, throughout the continent in the twentieth century was encouraged by the former colonial powers. Farmers were required to grow famine reserve crops as part of a district self-sufficiency policy in a period when poor communications made relief of famine difficult. Despite the compulsory measures associated with cassava, its agronomic advantages make it a popular and widely grown crop. Sweet potatoes are also popular where they can be grown, but are more specific in their ecological requirements.

Significant shifts in the relative proportions of crop areas have resulted from changes in food-crop preferences. An example, noteworthy for both the area involved and the relatively short period in which it occurred, was the replacement of sorghum and bulrush millet by maize as the major food crop in Sukumaland, Tanzania, between the mid-1950s and early 1960s (Peat and Brown 1962). However, Sukumaland is not well suited to maize production and in recent years a series of dry seasons has resulted in some shift back to millet and sorghum.

The introduction of maize, as both a food and cash crop, resulted in major changes in the farming system of the Plateau Tonga in Zambia. The traditional staple of the Plateau Tonga was sorghum, grown in mixture with other crops, including bulrush millet, groundnuts, cowpeas, and bambarra nuts. Following the example of European settlers, the Plateau Tonga began to introduce maize cultivation and the use of ox implements into their farming systems from the late 1920s onward. By 1933, there was already general adoption of maize cultivation on the more fertile soils. The Tonga adopted methods of cultivation which they saw used on the farms of European settlers. Mixed cropping and customary crop successions were gradually abandoned and the mixed sorghum garden replaced by large and poorly cultivated maize fields (Trapnell and Clothier 1937). By 1945, maize in pure stand occupied 80 to 90 percent of the acreage of Tonga farms. Sorghum had become a secondary crop, not grown by the larger farmers. The Plateau Tonga have a reputation as innovators. A survey carried out in 1967 indicated that while maize was still the dominant crop, grown on 63 percent of the cropped acreage of a sample of progressive farmers and on 70 percent of their neighbors' farms, groundnuts had become an important second crop as a result of a market becoming available. Cotton, too, was grown by some farmers (Anthony and Uchendu 1970).

The acceptance of a new crop by farmers is made easier when it fits readily into accepted farming systems. With the Sukuma and Tonga, and in other areas of tropical Africa where it has become an important crop, maize was readily substituted for other cereals. Similarly, where groundnuts have become a cash crop, there has been no problem in adjusting the area planted to it and other traditional crops. The spread of cassava has been helped by its suitability as the last crop in the rotation. In Teso District of Uganda, the introduction of cotton in the first decade of the century resulted in a modification of the traditional Teso farming system by making cotton, instead of finger millet, the opening crop in the cropping sequence. Land opened for cotton was easily made into a good seedbed for finger millet, sown the following year, and this led to easy acceptance of the cotton crop. But acceptance of a crop does not imply a major change in the priorities given to work allocation, and in Teso District, the main crop, finger millet, still has priority.

Factors affecting the rate of acceptance of farming innovations are discussed in later chapters. Profitability, ease of incorporation into the farming system, and low risk to the farmer are key factors. The importance of high return to investment and ease of adoption are highlighted by differences in response to maize and cotton in the southern province of Zambia. The net profits of cotton cultivation are very similar to those of maize. However, cotton is a troublesome crop: it needs frequent spraying to protect it against insect pests, is more labor-demanding than maize, and is difficult to adjust to the labor needs of other crops. Consequently, maize is the preferred cash crop of the area and development of the cotton crop has made slow progress. However, some farmers will continue to grow cotton, despite poor returns, because the crop will always give some return, even in a very bad year and when mismanaged.

High profitability can compensate for other drawbacks. For example, in Kenya, tea has been adopted with enthusiasm by small farmers because the high cost of establishment and the comparatively long waiting period from planting to first bearing was compensated for by the expectation of high financial returns.

In some areas the adoption of a cash crop has led to increasing specialization and a reduction in the amount of land and labor devoted to subsistence-crop production. An extreme example is the cocoa-growing areas of the Western State of Nigeria referred to in Chapter 2, where household consumption surveys in 1951/52 and in 1963/64 showed farmers to be purchasing from 46 to 77 percent of all calories they obtained from staple foods (Jones 1972:60). In the North Central State of Nigeria, a groundnut industry developed after World War I in response to the emergence of a world market for the crop. Previously, groundnuts were grown extensively throughout northern Nigeria, but they were grown in small quantities and were a neglected crop. Now the crop occupies 30 to 40 percent of cultivated land in some areas. This growth has been at the expense of food crops, and some places, as for example the areas in the vicinity of Katsina, have become net importers of grain.

Limited geographical specialization within the relatively small area of an administrative district has been recorded in the Kisii area of Kenya. At the turn of the century the major enterprises in the Kisii domestic economy were cattle owning and cultivation of food staples. The main staples were finger millet and sorghum, finger millet being the more important. Maize was grown, but the low-yielding, multicolored strain cultivated was not greatly valued. Major agricultural changes have taken place in the high-altitude areas of the district in the last twenty years. With the introduction of improved maize varieties, and latterly Kitale hybrid and synthetic varieties, maize has become the main cereal grown. The introduction of a highly profitable cash crop, pyrethrum, into Kisii District in 1952 stimulated the acceptance of other enterprises. Tea was introduced to small farmers in 1957 and was received with enthusiasm. Receipts from pyrethrum and tea provided capital for farmers to invest in pedigreed cattle, first made available to them in 1963. The introduction of high-quality dairy stock prompted farmers to make their holdings stockproof and to master new techniques of animal management. It was observed in 1967 that traditional food crops had become less important and farmers were beginning to look to those parts of the district that were unsuitable for pyrethrum and tea cultivation for supplementary food supplies (Uchendu and Anthony 1975b).

In areas specializing in tree-crop production, the cultivation of perennial and annual crops are usually distinct enterprises. But during the establishment of a plantation, the opportunity is often taken to grow food crops between the young trees. On the Kenya coast, annual cotton is sown into newly planted coconut and cashew and helps pay for the establishment of the plantations. In the forest areas of Ghana, the main farm enterprise is a system of fixed plantation cropping while food crops are usually grown in a natural fallow system. Land for food crops is in short supply and cocoa farming is so specialized in some areas that farmers who are able to produce all their food needs are not considered to be important cocoa planters. Farmers have become adjusted to the idea of depending upon the market for part of their food supply. In a sample of sixty farmers interviewed by Uchendu in the Akim-Abuakwa area of Ghana in 1967, nearly one-third had bought food in the preceding year and the cost of food per farmer varied from N₵92 to N₵251.[13] In response to the demand for food, specialized commercial food production was developing in some areas, for example the Begoro District, where cocoa farming was seriously limited by prevalence of swollen shoot disease (Uchendu 1969).

Standards of management on smallholder plantations are often indifferent and yields are correspondingly poor. Mixed cropping may be practiced, with attendant competition between the components of the mixture for soil nutrients and water, and difficulties in pest control. For example, coffee is commonly grown in mixtures with bananas and fruit trees and subsistence crops are interplanted where growth of the tree crops permits. A well-managed and productive plantation implies attention to such factors as satisfactory weed control, manuring, pruning, water conservation, pest control, and the removal and replanting of old unproductive trees. High standards of cultivation are, however, characteristic of outgrower schemes in which satellite smallholder plantations surround a nucleus estate, giving smallholders access to a mill and providing them with technical assistance, planting material, fertilizer, and credit. This has been the basis of smallholder oil palm development managed by SODEPALM in the Ivory Coast, and of smallholder tea development in Kenya. In Kenya, much of the smallholder tea is now a distance from estates and centered on factories in

[13]Or approximately the same number of U.S. dollars.

which the smallholder can buy shares. The ingredients for success have been a market for high-value produce, availability of inputs, and clear evidence of benefits from following the example of the large estates.

## Animal Draft Power

Work animals have been used in plowing and cultivating by small African farmers only within the last seventy years. Animal draft power is now used to varying degrees throughout savanna areas that are free of tsetse fly, but the possibilities provided by the use of ox-drawn implements for easing the pressure of labor shortages in the farming system are rarely fully exploited.

Teso District, Uganda, was one of the first areas in tropical Africa to adopt the ox plow, which was introduced to the district in 1910. The rapid acceptance of the plow was associated with the simultaneous spread of cotton as a cash crop and an abundance of animals. By the late 1960s practically every farmer used an ox plow for land preparation. Those farmers who did not possess a plow borrowed or hired one. Use of the plow resulted in larger areas under crops than would have been possible were only manual labor available. Farmers interviewed in Serere and Soroti counties in 1966/67 cultivated an average of 1.3 to 1.5 acres per person (Uchendu and Anthony 1975a). The two main crops, finger millet and cotton, are both labor-intensive and the availability of farm labor at the beginning of the crop season has become a limiting factor in production. Finger millet is planted first, either sown dry or at the beginning of the rains. Cotton sowing begins when sowing of the main food crops has been completed. Weeding finger millet competes for labor with cotton planting, and harvesting of finger millet with cotton weeding. Shortage of labor for weeding could be remedied by the use of ox-drawn seeders, for row planting, and of weeders, but no serious attempt was made to encourage the widespread use of implements other than the plow until the 1960s. By then, the enthusiasm generated by the first introduction of the plow and of cotton had lost its impetus, and progress has been slow.

The conditions under which the plow was adopted by the Plateau Tonga of Zambia are similar: Adoption was associated with the spread of a cash crop, maize, among a cattle-owning people with abundant animals and land. The Tonga were introduced to the plow at about the same time as the Iteso, but through contact with European settlers who used a wide range of ox-drawn implements. The Tonga learned by example, and their adoption of the plow was soon followed by the purchase of carts, seeders, harrows, and cultivators. Subsequently, the purchase of ox-drawn implements received a boost when credit was made available to Tonga farmers during the period of the African Improved Farming Scheme (1945–64). Most Plateau Tonga farmers now possess a range of ox-drawn implements which enable them to sow and maintain large areas in maize. Together with hybrid seed and fertilizer, the implements have contributed to the very substantial progress made in the area and the emergence of a class of medium-scale farmers, some of whom are now moving to tractor mechanization.

The amount of capital required to purchase trained oxen and equipment and the size of holdings needed to justify their ownership have led to slow adoption of ox-drawn equipment. The difficulties of adopting plow farming make it advisable to introduce profitable innovations simultaneously and to provide farmers with easy credit facilities. Many farms are too small, and sometimes too fragmented, for it to be worthwhile to invest in ox equipment. In an area of northern Katsina, Nigeria, surveyed by Anthony and Johnston, the minimum economic size of a plow farm was estimated to be about fifteen acres. The mean area under crops on a sample of farms on which there were no plows was six to seven acres, and on those with plows, twenty acres. The difference in terms of acreage cultivated per person, 1.0 and 1.7. acres respectively, was considerably less because households with plows tended to have more persons per household as well as a larger area. In two village areas, farms of fifteen acres or more made up 6 and 14 percent of the total number of farms. Helped by a scheme that provided credit for the purchase of oxen and implements, and with the opportunity for acquiring a cash income by groundnut cultivation, a high percentage of these farmers had plows. Assuming a cultivable percentage of 60, the maximum density of plow farmers that can be supported in this area of northern Katsina is about twenty-six per square mile. In 1967, the area supported fifty to sixty farms per square mile. Extension of ox plowing in this area is likely to result in the displacement of smaller farmers. This was found to be happening in the survey area (Anthony and Johnston 1968).

## Tractor Mechanization

Since World War II the use of tractors has frequently been regarded as the universal panacea for the problems of increasing farm productivity in Africa. Mechanization has often seemed the simplest and quickest way of modernizing traditional farming, and several costly programs for large-scale mechanization were put into effect, with little preparation, in the 1960s. These schemes follow an all too familiar pattern, beginning with the purchase of large quantities of expensive equipment and a crash training program to provide drivers and maintenance staff. Running costs were heavy and not accompanied by correspondingly large increases in production. Finally, schemes have been scaled down to manageable levels, more suitable for gaining experience on the introduction and management of tractor equipment.

Government tractor-hire services have been provided for small-scale farmers in Africa for at least twenty years. Their object has been to help farmers achieve timely cultivations. Unfortunately, demand has largely focused on land preparation. In Uganda in the late 1960s, approximately 60 percent of the hours worked by the tractor-hire service was spent plowing (Brown, Evans-Jones, and Innes 1970). Partial mechanization, which only helps farmers to sow larger areas of crops, is likely to aggravate labor shortages later in the season when operations are carried out manually.

To obtain an efficient sequence of operations, mechanization must alleviate and not accentuate labor shortages in the farming system. Selective mechanization has been used to advantage in the rice irrigation scheme at Mwea/Tebere in Kenya. In the early days of the scheme, land preparation for paddy rice was carried out with work oxen. It was difficult to complete the operation on time over a large area and the timetable for other operations suffered accordingly. The subsequent introduction of tractor-drawn equipment made for more thorough land preparation, and it became possible to ensure that planting and field operations were carried out. Firm control by management of the scheme's tractor-hire service contributed to the success of the scheme.

Tractor-hire services have been costly and charges to farmers are frequently heavily subsidized. H. Mettrick reports a charge of 18.23s. an hour in Uganda in 1966 for tractors that cost 35.94s. an hour to operate, including depreciation (Mettrick 1970). High costs result partly from the supply of services on a national basis to small, dispersed fields, and from very seasonal work. Most of the demand for tractor services in Uganda fell in a four-month period; tractors owned by the tractor-hire service worked on average only 500 hours per year, compared with the 1,000 to 1,200 hours required to keep costs at a reasonable figure (Brown, Evans-Jones, and Innes 1970). The provision of service facilities to small dispersed units adds to maintenance and repair costs, as does the operation of tractor-hire services within the administrative framework of the civil service.

Private tractor owners have been more successful, and often provide a cheap and efficient service. Costs are kept down by working long hours, by the greater care taken of equipment, and by the greater freedom owners have to choose their customers. It has proved advantageous to leave development of tractor-hire services in the hands of entrepreneurs, who can call on government extension services for advice on the types of equipment to use.

Faced with the problem of economic operation of tractors in a farming system where fields and profits are small, governments have turned to the creation of large blocks of land, made up of the holdings of a number of participating farmers and with operations usually directed by a farm manager. Group farms in Uganda and the block farms of Tanzania are examples. Practices differed, but land was usually prepared for planting by the management, while settlers were responsible for other operations including harvesting. Spraying was by plane or was the responsibility of the farmer.

Group farming has experienced a high rate of failure. Avoidable technical causes, as, for example, faulty siting and inadequate support facilities, were often to blame and reflected the haste with which the schemes were established. An important reason for the poor performance of group-farming projects in both Uganda and Tanzania was the conflict between the requirements of a participant's private family farm and his new fields in the group units, usually to the detriment of the latter. Farmers lacked the incentive to cooperate, and crop returns rarely justified the increased cost of tractor use. Only a few farmers have been affected by these units. Their principal consequence has been the diversion of scarce resources of capital and extension staff from projects that could have had greater economic impact, such as extending the range of ox-drawn equipment used by farmers.

## Government Intervention

Government programs have had a major influence on farming methods and systems. But, with a few notable exceptions, policies aimed at the direct transformation of traditional farming systems have made less impact than measures that have encouraged changes in the components of the system. Farmer response to new crops and techniques has been influenced most by policies that have affected market opportunity and the price paid for produce, the opportunities presented to farmers for increasing the return to investments, and the availability of credit and inputs. These have all been identified as stimuli to change in the Mazabuka District of Zambia. The readiness of farmers in Mazabuka District to change their farming systems over the last sixty years has resulted from a combination of factors, especially the suitability of the area for maize cultivation and the example provided by the success of European settlers. The acceptance of new farming methods was further stimulated by government programs which resulted in the provision of easy credit and a network of rural markets and ensured the availability of improved seed, fertilizers, insecticides, and other requisites to the small farmer. Between 1946 and 1962 the government's African Improved Farming Scheme provided capital for farmers who followed a prescribed rotation, maintained a satisfactory standard of cultivation, and made use of fertilizer and manure. The achievements fell short of original hopes, and the scheme has been criticized because its benefits were confined to a limited number of farmers in the southern and central provinces located near the line of rail (Makings 1966:216–23; Bates 1976:38). Moreover, crop yields remained low during the period of the scheme, and it was expensive in terms of the manpower required for supervision and numbers of participating farmers. Nor did it lead to the planned integration of crops and livestock. However, the injection of capital helped farmers to increase their investment in implements and improvement of their holdings and made them more ready to adapt other innovations as they became available (Anthony and Uchendu 1970).

Government action has been most effective in bringing about change where it has been associated with measures that helped farmers to accept a visibly profitable innovation or sequence of innovations. In Mazabuka District, the successful adoption of maize as a cash crop and of ox-drawn implements influenced farmers to adopt other successful innovations, including hybrid maize seed, fertilizer, and cotton cultivation.

The mixed farming campaign in northern Nigeria was another major attempt by government to help farmers modify their farming systems. The aim was the widespread integration of crop and animal management. A system was developed whereby a farmer with two oxen could plow and maintain the fertility of an area large enough to provide his family with their subsistence needs and give a crop surplus for sale. Participating farmers were provided with credit to purchase trained oxen and equipment and were helped to train their animals. Farmers had to make a pen in which to house the oxen, and were expected to supply fodder and bedding grown on the farm. Pen manure was collected and applied to a portion of their farm each year. Starting in 1928, the scheme grew slowly during the period when the returns to be had from the two cash crops in northern Nigeria, groundnuts and cotton, were small. However, the mixed farming scheme continues to make progress and after forty years there are somewhat more than 40,000 mixed farms in the northern states of Nigeria.

Yet the scheme has been only a partial success. A minority of farmers have been affected, and a survey of farmers in northern Katsina indicated that, because of the work involved, few of those farmers who had adopted ox plowing still observe the scheme's rules for the preparation of pen manure. However, the scheme has helped to establish the ox plow in the savanna areas of Nigeria, and further adoption of the plow by farmers is limited more by availability of capital and land than by any lack of ambition to become a plow farmer. Plow farmers have emerged as a distinct class. They usually have larger farms than their neighbors and are more prosperous.

The African Improved Farming Scheme of Zambia and the Mixed Farming Scheme of Nigeria provided individual farmers with incentives to change their farming methods. In contrast, the *paysannats indigènes* of the former Belgian Congo involved whole communities in a countrywide resettlement program. The *paysannats* system was an attempt to rationalize existing farming systems. It retained the traditional short cropping period and long fallow. In the forest area of the Congo, farmland was laid out in long fields or corridors, running from east to west and each 100 meters wide, at right angles to the farm holding. The length of a strip depended upon the number of farmers settled and the area of individual plots, and one strip was allocated to each phase of the rotation. For example, where three years of cropping were followed by twelve years of forest fallow, fifteen corridors were established. Clearing was carried out successively on alternate corridors, so that each cropped field was bounded by well-established fallow or forest, in order to help regeneration after land was returned to fallow. In savanna areas, where corridors were not considered to be essential, the land was often laid out in blocks which were divided among family holdings, or *fermettes*. Part of each holding was allocated to a homestead area for buildings and kitchen garden and the rest divided into a series of rectangular fields allocated to the cultivation cycle. At independence, nearly half a million families had been settled in *paysannats* and *fermettes*.

The objective of this settlement program was to stabilize the population and to provide a suitable environment for the introduction of improved seed and implements and better farming practices. It was intended that the system would be intensified as knowledge was obtained on how to do so. However, it had notable shortcomings (Dumont 1957; Allan 1965). The geometrical layout of farms took little account of soil fertility and terrain.[14] The scheme was too rigid, and a given area of crops was expected to be cultivated by all families irrespective of their size. In the forest areas, this often resulted in the laborious work of clearing very tall vegetation from

[14]One side effect of the land surveys was to bring under cultivation land that lay in unoccupied "buffer zones" just outside a village's customary boundaries.

areas which, without compulsion, a small family would not have cultivated.

Little information is available on which to judge the progress of the *paysannats* in the years following independence. Their continued development was dependent on close supervision, a flow of production-increasing techniques, and an efficient marketing system. It is probable that these have been difficult for government to supply. In its original form, the *paysannats* system was dependent on an abundant supply of land and was not designed for areas where population density no longer permits farmers to follow their traditional fallow system.

The Azande Scheme, begun in the southwest Sudan in 1945, was contemporary with the *paysannats* and had a similar settlement pattern. The Azande were settled in dispersed villages made up of about fifty families. Each family was given between twenty and thirty acres of land in a long rectangular plot. Resettlement was intended to relieve the overcrowding of population that occurred along the roads, where the Azande had earlier been moved for sleeping sickness control. Cotton was to be the main cash crop and to provide the raw material for a spinning and weaving mill. Grey cloth would be the major export from the district. Other industries would include an oil mill, a soap factory using locally produced cottonseed and palm oil, and a jaggery factory for processing sugarcane.

Within each holding the farm family was allowed to practice traditional farming methods, with the limitation that the new holdings often did not contain a full range of soil types which the Azande were accustomed to use. The initial enthusiasm of the Azande for resettlement and cash-crop production waned as cotton was found to be a labor-demanding crop giving comparatively poor returns. Even on the Yambio Experimental Farm, the agricultural research station serving the area, yields were not high. The labor needs of the cotton crop competed with those for food crops, and cotton picking and end-of-season stalk destruction, recommended as a pest-control measure, clashed with traditional dry-season activities—hunting and wild honey collection. The Azande gradually became disillusioned with change and the scheme did not survive the period of unrest that began in the southern Sudan in 1955.[15]

Direct government intervention to establish a new farming system has been most successful where a favorable environment has permitted the operation of a successful farm enterprise for the benefit of individual farmers. The Gezira Scheme in the Sudan is a classic example. The main

[15]See C. C. Reining (1966) for a critical account of the Zande Scheme.

ingredients of success were the high potential of Gezira soil, a suitable environment for growing extra-long-staple cotton, a terrain suitable for a large-scale irrigation scheme, and a partnership between tenant farmer, management, and government that provided the farmers with all the benefits of participation in a large-scale enterprise. The Gezira is now a major world supplier of Egyptian-type cotton.[16] Another successful, if small-scale, irrigation scheme with government participation, mentioned earlier in the chapter, is the Mwea Rice Scheme in Kenya. It has prospered because of the suitability of the heavy black soils of the area for irrigated rice production and the high price obtainable for the crop locally.

The importance of profitability to farmer response to change is a recurring theme throughout this book. The readiness with which farmers accept economically worthwhile changes to their farming systems gives cause for optimism for the future. The experience of Kenya shows that, given suitable circumstances, major changes can be achieved comparatively rapidly. Two programs have made a major impact in Kenya—the land consolidation and the land settlement programs.

By the 1950s a lot was known about the profitable management of land of high potential in Kenya, as a result both of research and practical experience. Unfortunately, excessive fragmentation of farmland made it impossible to put sound development plans into effect in many parts of the country. Fragmentation was most severe in heavily populated districts such as Nyeri, Kiambu, and Maragoli, but efforts to obtain voluntary consolidation had met with little success (Brown 1957). However, acceptance of the Swynnerton Plan by the Kenya Government in 1954 enabled a program of land consolidation to be started. The aim of the Swynnerton Plan was to provide African farmers with security of land tenure on a farm capable of providing a family income comparable with that earned in other occupations. The state of emergency then prevailing in Kenya had led to recognition, at all levels, of the urgency of solving the country's agrarian problems, and was a major factor in securing its implementation. Under the plan, staff and funds were made available for further intensification of agriculture, including the tasks of mapping, consolidation, adjudication, and registration of farmland. Consolidation has been accompanied by en-

[16]In the 1960s, a program of crop intensification and diversification was begun in the Gezira. In the Main Gezira (excluding the Managil Extension), there was an increase in cropping from less than 45 percent in the early 1960s to more than 60 percent of the gross rotational area in the early 1970s. Cotton still accounts for 25 percent of the gross rotational area in the Gezira, a proportion that has remained unaltered since the Gezira eight-course rotation was adopted in 1933 (Farbrother 1973). However, there have been notable increases in the areas of other crops, particularly of wheat and groundnuts.

closure, and this has contributed to a rapid expansion in dairy production based on "grade cattle," i.e., crosses between high-yielding European breeds and indigenous Zebu cattle. The provision of a title to his land makes it easier for a farmer to secure additional credit, encourages him to improve his land, and reduces the amount of litigation over land ownership. By the end of 1965, 1.6 million acres had been registered, and the report of the Lawrance Mission on Land Consolidation and Registration in Kenya proposed an accelerated program (Lawrance 1967). By 1972/73 some 4 million acres in agricultural areas and another 2 million acres in range areas had been brought into the land register in approximately 630,000 holdings (Kenya 1974:218).

The Kenya Government has continued to give a high priority to land consolidation and registration. This is not an end in itself but rather a program that facilitates follow-up programs for agricultural education, credit, and other measures to promote expanded production (Lawrance 1967). However, others are skeptical about the benefits resulting from the large investment in land consolidation, enclosure, and registration. Thus Judith Heyer argues that the increases in production in the 1950s and 1960s are mainly the result of the relaxation of the restrictions which had prevented African farmers from growing cash crops, especially coffee, and from keeping dairy cattle. In her view there is little evidence that they "were attributable in any way to the land reform programme" (Heyer et al. 1976:11). On the basis of a detailed analysis of the impact of the efforts to introduce individual tenure on agricultural planning, farmer decision-making, credit, and on the pattern of land distribution, H. W. O. Okoth-Ogendo (1976) concludes that the adverse effects are probably more significant than the expected benefits. But, as he notes, it is difficult to evaluate the impact of tenure reform per se on agricultural development because of the simultaneous influence of many other factors ranging from new roads and schools to the spread of new technologies. It also seems likely that the impacts have differed considerably from region to region. Both the positive effects in fostering expanded output and the negative effects in contributing to increased landlessness may well have been more pronounced in Central Province than in other regions because of the potential for coercion created by the emergency.

A program of land purchase from European farmers in Kenya for small-holder resettlement has involved another 1.2 million acres. The scheme, which had important political implications, was undertaken to reduce land pressure in overpopulated areas and to provide a means of introducing more sophisticated and intensive farming methods to large numbers of farmers (Maina and MacArthur 1970). The second objective applied particularly to the so-called "low-density scheme" which was "designed for those Africans who had demonstrated agricultural ability and could raise some capital on their own . . ." (Senga 1976:84). The great majority of the 35,000 families that acquired land under the Million Acre Settlement Scheme were part of a "high-density scheme," and were allotted much smaller holdings which permitted production of only a modest marketed surplus above family requirements. There have been serious loan repayment problems with both types of settlers, but the scheme did maintain production in the Highlands. It also provided some relief of population pressure in densely populated farming areas, although the £30 million cost of the scheme absorbed a large percentage of the funds available for agricultural development.

## The Future

Traditional farming systems have changed and are changing in response to changing conditions. Much of the change has been unplanned. It has come about through the opening of new market opportunities for the products of the land and through land pressure resulting from an increase in population and livestock numbers and the land requirements of the new cash crops. Movement of population, the traditional method of easing local population pressure, was restricted by the colonial powers. Shifting cultivation has been largely abandoned in favor of fixed homesteads and natural fallow systems, and shortages of land have forced farmers to modify their farming systems further by shortening the fallow period and sometimes by shifting to crops that are less demanding of labor and soil nutrients. If the fallow is made too short, the soil loses fertility and crop yields decline. Intensification of cropping, without modification of farming methods, will eventually result in a decline of agricultural production. That is the technical challenge to which changes in African farming systems must respond.

# 5. The Varied Pattern of Achievement as Revealed by the Field Studies

The seven field studies undertaken in connection with this study were designed to sample farming systems with varying levels of economic achievement, and to attempt to determine the circumstances that were associated with success and failure. The areas differed in resource endowment, economic environment, and the nature of government assistance. Their economic and political histories were also quite different, but all had been under British rule from about the 1880s through the 1950s. Each had experienced some changes in farm organization within the last few decades, but only in a few had this resulted in sustained economic growth.

This chapter explores the achievements of the eight agricultural systems represented in the study areas and the observed conditions that might have been expected to affect them. Some attention is also devoted to the process of change in each area prior to the time of the study.

## The Study Areas

The location of the study areas and the major bioclimatic regions (according to John Phillips) are shown on Map 5.1.[1] Five of the study areas are in savanna zones, but Akim-Abuakwa lies in the secondary rain forest that is characteristic of much of coastal West Africa and Kisii is a mountainous region in which moist montane forest has been replaced by cultivated savanna.

Table 5.1 presents relevant information about the study areas. It tells something about the resource endowment of the areas, the infrastructure, and the innovativeness of farmers. Intrinsic soil fertility and potential yield of cash and food crops may be taken as measures of resource endowment; extension coverage and distance to a railway as proxies for infrastructure; and land enclosures, land markets, and ox-drawn plows and other ox-drawn implements as indicators of innovativeness. Table 5-2 shows the

[1]Phillips's bioclimatic regions are intended to identify agricultural potential in terms of ecological potential.

*p 5.1*. Location of field study areas and major bioclimatic regions*

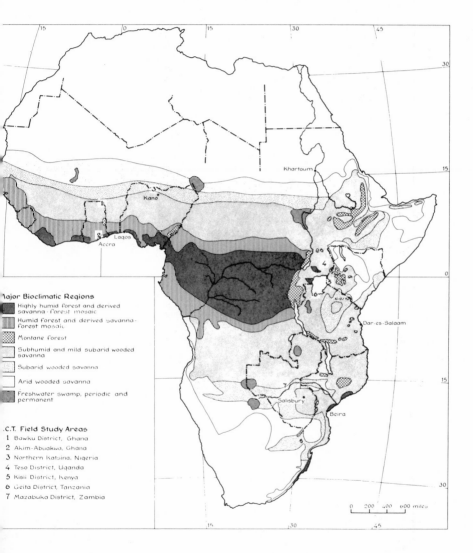

**Major Bioclimatic Regions**

- Highly humid forest and derived savanna-forest mosaic
- Humid forest and derived savanna-forest mosaic
- Montane forest
- Subhumid and mild subarid wooded savanna
- Subarid wooded savanna
- Arid wooded savanna
- Freshwater swamp, periodic and permanent

**C.T. Field Study Areas**

1. Bawku District, Ghana
2. Akim-Abuakwa, Ghana
3. Northern Katsina, Nigeria
4. Teso District, Uganda
5. Kisii District, Kenya
6. Geita District, Tanzania
7. Mazabuka District, Zambia

*Bioclimatic regions based on John Phillips, "Map of Africa South of the Sahara: Major Bioclimates," in *Agriculture and Ecology in Africa*, by John Phillips (London, 1959).

Table 5.1. Characteristics of the field study areas*

| Study area | Katsina | Mazabuka | Geita | Bawku | Teso | Akim-Abuakwa | Wanjare (Kisii) | East Kitutu (Kisii) |
|---|---|---|---|---|---|---|---|---|
| Country | Nigeria | Zambia | Tanzania | Ghana | Uganda | Ghana | Kenya | |
| Latitude | 13°N | 16°S | 2°S | 10°N | 1°N | 6°N | 0° | |
| Mean altitude (feet) | 1,600 | 3,500 | 3,800 | 700 | 3,500 | 300 | 5,000 | 6,500 |
| Mean annual rainfall (inches) | 27 | 33 | 37 | 40 | 54 | 60 | | 70 |
| Months of rain | 7 | 4.5 | 6 | 7 | 8 | 9 | | 12 |
| Unimodal or Bimodal distribution | U | U | U | U | B | B | | B |
| Intrinsic soil fertility | low | fair-high | mod. | low | mod. | good | | high |
| Percent of land cultivable | 60 | 25 | n.a. | 50 | 70 | 70 | | 90 |
| Persons per square mile | 200 | 50 | 100 | 150 | 130 | 150 | 500 | 900 |
| Cattle per 100 persons | 25 | 145 | 50^a | 20 | 120 | 0 | | 40 |
| Mean farm size (acres) | 8 | 16^b | 25^c | 10 | 20 | 25^c | | 8 |
| Number of blocks per farm | 3 | 1–2 | 1 | 3–4 | 1–2 | 5 | | 1 |
| Enclosure | neg. | low | low | nil | neg. | nil | | high |
| Land market development | high | low | nil | nil | mod. | high | | good |
| Main cash crop | peanuts | maize | cotton | peanuts | cotton | cocoa | coffee | pyrethrum, tea |
| Staple food crops | millet, sorghum | maize | maize, manioc | millet, sorghum | finger millet, sorghum | cocoyam, plantain | maize, sorghum, finger millet | |
| Potential yield of cash crops | mod. | high | good | fair | good | high | fair | high |
| Potential yield of food crops | fair | high | mod. | low | fair | fair | mod. | good^e |
| Extension coverage | low | good^d | low | low | low | low | | neg. |
| Adoption of ox plow | fair | high | neg. | low | high | nil | | nil |
| Adoption of other ox implements | neg. | high | nil | nil | low | nil | | nil |
| Mean miles to railhead | 100 | 0 | 30 | 400 | 0 | 15–20 | | 30 |

*Qualitative assessments are nil, neg. (negligible), low, fair, mod. (moderate), good, high, n.a. (not available).
^a Estimated.
^b Most farmers cultivate 8 to 15 acres, but the range in farm size is large and a few farmers cultivate more than 100 acres.
^c Farmers may have extensive holdings outside the district.
^d Extension coverage for cotton growing was high.
^e Extension coverage for tea growing was high.

*Table 5.2.* Field study areas ranked according to relevant characteristics
(on a scale of 1 to 5)

| Study area | Natural potential | Infrastructure | Innovation |
|------------|-------------------|----------------|------------|
| Mazabuka | 1–2 | 1–1.5 | 1 |
| Kisii | 1.7–2.7 | 2 | 2 |
| Akim-Abuakwa | 2.3 | 3.5 | [1][a] |
| Geita | 2.7 | 3.5 | 3.5 |
| Teso | 3 | 3 | 2.7 |
| Katsina | 4 | 4 | 2.7 |
| Bawku | 4.7 | 5 | 4.5 |

[a]Land market score only.

rankings that result when each of these characteristics is scored on a scale of 5, with 1 high and 5 low, and scores averaged for each of the categories. The range of scores for natural resources and infrastructure is greater than for innovativeness. Either farmers in the seven areas are more or less equally innovative, or the measures chosen are poor indicators. In fact, each is biased either by cultural or environmental factors: land markets are much more highly developed in West Africa than East, and enclosure and ox-drawn implements are only likely to be found in areas where cattle are kept. Nevertheless, the measures of innovativeness, as well as those of resources and infrastructure, may help to explain differing economic achievement.

Comparison among the areas is facilitated when two or more grow the same crop. Cotton is a major cash crop in three areas, peanuts in two, and tree crops in both of the high-rainfall areas. Maize, sorghum, and millet are found in various areas. Cattle are present in all areas except Akim-Abuakwa, and there is some use of oxen as draft animals in each area where cattle are present. Akim-Abuakwa is also unique among the study areas in its principal reliance on starchy roots and bananas for food crops.

## Productivity

Table 5.3 and Chart 5.1 present very rough calculations of the value of average gross and net cash earnings and expenditures for "progressive" farmers and "neighbors" in each of the study areas. In Mazabuka, Kisii, and Teso the terms "progressive," "improved," or "emerging" farmer were applied to farmers who had satisfied certain requirements of the

Table 5.3. Gross and net cash income of progressive farmers and neighbors (in shillings)*

|  | Mazabuka | Akim-[a,b] Abuakwa | Kisii E. Kitutu | Teso | Katsina[c] | Geita | Kisii Wanjare | Bawku[a] |
|---|---|---|---|---|---|---|---|---|
| **Per farm** | | | | | | | | |
| **Progressive farmers** | | | | | | | | |
| Gross cash income | 9,453 | 6,657 | 2,924 | 1,588 | 1,547 | 1,344 | 1,108 | 469 |
| Purchased inputs | 4,382 | 2,099 | 1,275 | 378 | 903 | 196 | 720 | 92 |
| Net cash income | 5,071 | 4,558 | 1,649 | 1,210 | 664 | 1,148 | 388 | 377 |
| **Neighbors** | | | | | | | | |
| Gross cash income | 1,977 | 1,533 | 1,695 | 809 | 419 | 540 | 256 | 90 |
| Purchased inputs | 907 | 622 | 483 | 141 | 190 | 90 | 86 | 14 |
| Net cash income | 1,070 | 911 | 1,212 | 668 | 229 | 450 | 170 | 76 |
| **Per acre cropped** | | | | | | | | |
| **Progressive farmers** | | | | | | | | |
| Gross cash income | 284 | 122 | 209 | 98 | 86 | 58 | 112 | 31 |
| Purchased inputs | 132 | 39 | 91 | 23 | 50 | 8 | 72 | 6 |
| Net cash income | 152 | 84 | 118 | 75 | 36 | 50 | 40 | 25 |
| **Neighbors** | | | | | | | | |
| Gross cash income | 124 | 33 | 270 | 75 | 38 | 28 | 33 | 10 |
| Purchased inputs | 57 | 13 | 77 | 15 | 17 | 5 | 11 | 2 |
| Net cash income | 67 | 20 | 193 | 60 | 21 | 23 | 22 | 8 |
| **Farm size (acres)** | | | | | | | | |
| Progressive farmers | 56 | 54 | 14 | 34 | 20 | 23 | 10 | 15 |
| Neighbors | 21 | 46 | 6 | 13 | 12 | 20 | 8 | 9 |

*Data from Food Research Institute field studies.

[a]Converted from new cedis to shillings at the official rate of N₡1.00 = 7.14s., which was somewhat above the black market rate.

[b]Figures exclude one farmer's income of N₡3,600 (25,200s.) from the sale of cabbage.

[c]Includes off-farm cash income amounting to about 105s. per farm.

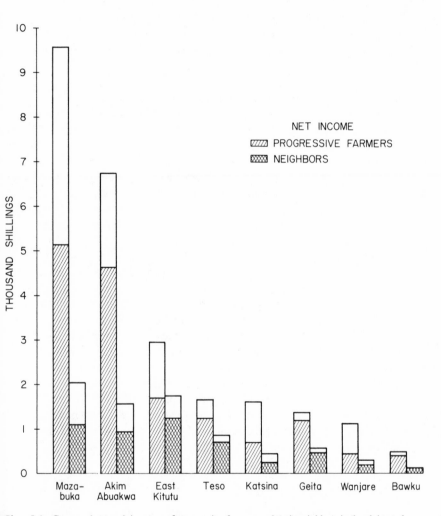

*Chart 5.1.* Gross and net cash incomes of progressive farmers and their neighbors in the eight study areas

Department of Agriculture as regards training or practices that qualify them to receive special concessions in access to knowledge, productive inputs, and credit. Progressive farmers are "better" farmers not in the sense that they produce more or earn more but that they follow the advice of extension workers.[2] In the other study areas, farmers designated as "progressive" in this chapter are simply those fifteen in each sample of sixty who had the largest gross cash income. Neighbors are all other farmers in each sample.[3]

The variation in net cash income per farm ranges from 76s. to 5,071s. and net cash return per cropped acre slightly less: from 8s. to 193s. Even if the comparison is made only between groups of farmers who occupy similar relative economic positions in their area, the range is large: from 377s. to 5,071s. for progressive farmers, and 76s. to 1,212s. for neighbors. Total farm income tends to be correlated with farm size, but the difference among progressive farmers is only from fifteen acres per farm in Bawku to fifty-six acres in Mazabuka, and neighbors' farms ranged from nine acres in Bawku to forty-six in Akim-Abuakwa. Size of farm more or less accounts for the much larger incomes in Mazabuka and Akim-Abuakwa, although returns per acre are on the high side for progressive farmers in both areas.

The variation in gross and net returns per acre, although smaller than variation in returns per farm, is still large. Net cash returns of progressive farmers in Mazabuka and East Kitutu are two to three times those of Geita progressive farmers, five to six times those of Bawku progressive farmers. Neighbors in Mazabuka and Teso net three times as much per acre as Geita neighbors do, seven and one half to eight times as much as Bawku neighbors, while East Kitutu neighbors net more per acre than progressive farmers. Measures of gross product per acre are similarly variable.

Whether attention is to be directed at net output (income) per farm, net output per acre, or gross output per acre depends on the purposes of the investigation. If welfare of the rural resident is the primary concern, the appropriate unit is the farm. If, on the other hand, the sector and its

[2]Similar classification of Rhodesian farmers as skilled, semiskilled, and unskilled farmers proved to be poorly correlated with farmers' achievements (Massell and Johnson 1968:60–61).

[3]The classification of progressive farms used in this chapter differs from that used in selecting the original sample (see Chapter 1, "The Field Studies of Agricultural Change"). The very small difference in net incomes of progressive farmers and neighbors in East Kitutu suggests that "progressiveness" may be correlated more with willingness to invest in new inputs than with net productivity. East Kitutu observations appear to be aberrant in various of the comparisons presented in Chart 5.1, and this may on occasion distort overall relationships.

contribution to national product are at issue, net value per acre may be more important. Gross output per acre would be the relevant magnitude only if food were valued above all other things, including labor, or if the concern were purely with technical rather than economic productivity.

A search for the causes of these large differences in farm incomes will help to identify the major determinants of agricultural change. The factors that might be expected to explain the variation among areas include the natural resource endowment; the state of the arts, both those that are practiced and those that could be; and access to technical knowledge, to product markets and to purchased inputs, and to land and labor. Differences in cultural values, social organization, and the political-economic order affect these explanatory variables; they may also affect the motivation and capacity of farmers to achieve economic goals.

Variation among farmer behavior within an area may result from many of the same causes that bring about variation among areas, from aspects of society that affect their access to markets, productive resources, and knowledge, and from noneconomic attributes such as race, lineage, religion, or party affiliation.

The field studies were designed to capture the effect of factors that might be expected to influence productivity both within an area and among areas. Some of the principal determinants that were examined, and the results of this examination, are discussed in the following section.

## Possible Causes of Varying Productivity

The circumstances that may cause productivity to vary from farm to farm and from area to area will be considered under six headings: resource endowment, state of the arts, access to markets and purchased inputs, diffusion of innovations, variation among farmers, and access to land and labor. The relationships between productivity and a number of these determinants in the various areas are shown in the panels of Chart 5.2. In the last section, the qualitative evaluations of the achievements in each area by Anthony, Johnston, and Uchendu will be reviewed.

### Resource Endowment

Under the heading of resource endowment may be grouped soil quality, climate, latitude, density of human and animal population, and incidence of parasites and disease. Measures of several of these are given in Table 5.1.

*Soil and climate.* The indications of soil and climatic quality given in

*Chart 5.2.* Relationships between various characteristics of farming in the study areas*

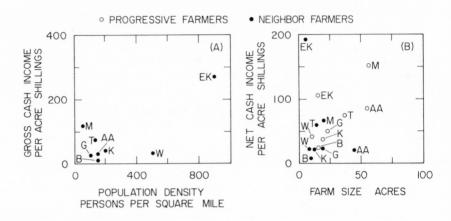

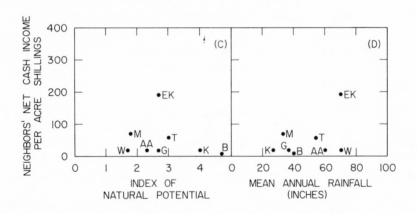

*Key: Initials identify study area: AA, Akim-Abuakwa, Ghana; B, Bawku, Ghana; EK, East Kitutu, Kenya; G, Geita, Tanzania; K, Katsina, Nigeria; M, Mazabuka, Zambia; T, Teso, Uganda; and W, Wanjare, Kenya.

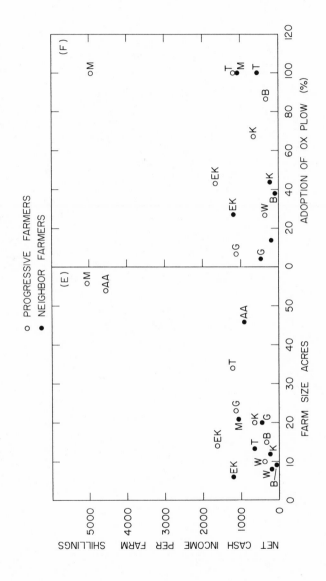

*Chart 5.2.* (cont.)

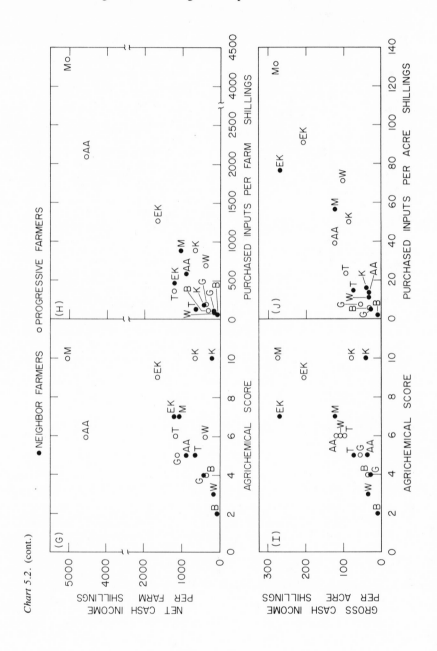

*Chart 5.2.* (cont.)

appraisals of intrinsic soil fertility and of yield potential should probably be discounted somewhat because they could not help but be influenced by observed productivity per acre on the better farms. If these appraisals are reliable, they might be expected to correlate fairly well with net return per acre on the neighbors' farms, where soil-improving practices and purchased inputs are least (Chart 5.2C). They do not.

The index of natural potential as calculated includes rainfall only indirectly as it affects yield potential, and it is difficult to include it directly. It is not simply a matter of more or less rainfall, although there is some tendency for total annual rainfall and return per acre to vary together (Chart 5.2D). The more important consideration is that in tropical areas where mean annual rainfall is as low as it is in Mazabuka, Geita, Bawku, and Katsina, and where there are marked wet and dry seasons, rainfall is likely to vary widely from year to year in total amount and in timing in ways that cause crop yields to fluctuate widely. In the years when these studies were conducted and the estimates of net yields per acre obtained, however, rainfall did not depart significantly from "normal" timing and amount except in Kisii, where deficient rainfall in 1966 caused the loss of the main maize crop.

*Latitude.* The study areas vary from 13°N to 16°S. The range is great enough to have a profound effect on rainfall, but not sufficient to have much impact on insolation or day length. Altitude is a critical determinant in the farming pattern of Wanjare and East Kitutu but only insofar as it determines the kind of crops grown.

*Density of human and animal population.* Population density enters as it measures total land resources relative to requirements and as it affects differential access to land. It might be expected to correlate positively with gross output per acre (Chart 5.2A). Again, the fit is poor. The ratio of population to land may be manifested in size of farm, but comparison of income with farm size is little more revealing (Chart 5.2E). Except for the large farms and large incomes of progressive farmers in Mazabuka and Akim-Abuakwa, there is little relation. If outliers are disregarded, there does seem to be some tendency for income to rise as farm size increases from ten to twenty acres.

*Parasites and disease.* Information about the incidence of parasites that might affect the productivity of men and animals is not presented in the reports on the field studies because differences between one area and another are small. Malaria is endemic in all areas, and intestinal and other parasites are widespread. Most of the study areas lie outside the tsetse-fly zone, but the fly has only recently been cleared from Geita, and nagana still makes Akim-Abuakwa unsafe for cattle.

## State of the Arts

In considering the technical achievements of African agriculture, attention should be directed both at the farmers themselves—the technology they pursue—and at the technical changes recommended by the extension service. Farmers' failure to adopt a particular technical innovation may result from ignorance of its potential or from recognition that increased returns resulting from its adoption are likely to be less than increased costs, or they may reject it for other reasons. On the other hand, adoption does not necessarily indicate that the farmer clearly and correctly perceives the economic advantages to be gained by the new implement or practice. Instead, he may be influenced by perceptions of status or by desire to conform to the extension officer's values. Adoption of a proffered innovation, like its rejection, may be determined by noneconomic as well as economic considerations.

*Recommended practices.*   A list of the kinds of practices extension services can offer to farmers would not vary much from one study area to another. It would undoubtedly contain some sort of an entry under each of the following categories:

> Improved varieties of crops now grown.
> Application of manure or chemical fertilizers.
> Application of pesticides to seeds before planting, to the growing crops, or to crops in storage.
> Improved cultural practices referring primarily to time of planting, plant population, and row cultivation.
> Use of draft oxen or of tractors.

Some lists might also include chemical herbicides; implements like peanut-hullers, corn-shellers, and corn-grinders for postharvest operations; and new crops, like tea in East Kitutu, as well as new varieties of familiar crops.

Within these general categories, however, there are differences in number and in quality.

Table 5.4 presents a generalized list of recommended practices for the eight study areas and indicates as well to what extent farmers knew about them and practiced them.

Adoption of recommended practices must depend both on the suitability of the innovations and on the effectiveness of the extension services and private suppliers in reaching farmers. In most of the areas studied, agricultural extension was the principal instrument for informing farmers of the advantages of new inputs and techniques, and the coverage of extension services varied greatly (Table 5.5).

The rate of adoption of technical innovations among the districts was correlated with the number of profitable ones that were available. Both Akim-Abuakwa and Teso districts seemed to have reached a plateau in their technical change some years ago; when innovations were profitable they were adopted, but at the time of the field studies not much was happening. On the other hand, Kisii and Mazabuka showed what can happen when economic conditions provide adequate incentives. Bawku District is handicapped by high marketing costs and neglect and looked like a forgotten area. Modest receptivity to change characterized farmers in Katsina, while Geita farmers showed eager response to profitable new opportunities. Farmer response to technical innovations in each district is summarized on the following pages.

*Mazabuka.* The important technical recommendations at farm level were centered on three crops: cotton, maize, and peanuts. The extension staff was strong and farmers were keenly aware of both cotton and maize recommendations. In general, the progressive farmers were the best-informed group.

Farmers who grew cotton knew the right time to sow, the spraying and fertilizer recommendations, and the names of the three insecticides used. Their knowledge of current agricultural recommendations was a testimony to the quality and the coverage of the extension staff, and to the economic soundness of their recommendations.

*Akim-Abuakwa.* The commercial production of cocoa in Ghana was pioneered in Akim-Abuakwa, and most of the agricultural innovations center on that crop. Cocoa farmers tended to have the recommended equipment for cocoa farming. More than 65 percent of the sample farmers owned and used motor-driven sprayers. In general, farmers' awareness of the technical recommendations for cocoa farming was good, but it had not always led to the adoption of improved practices. Regular spraying was limited by lack of insecticides and the employment of laborers who work for a share of the crop.

Knowledge about cocoa varieties was good. Eighty-six percent of the farmers knew of the superiority of Amazonia over Amelonado and 40 percent knew about the advantages of hybrid cocoa. Only 28 percent, however, were aware of any association between the swollen-shoot virus and the presence of kola or silk cotton trees.

Food-crop farming had not been changed by any major innovations except for one farmer who specialized in the production of cabbages for the Accra market.

*Kisii.* Farmers in both Wanjare and East Kitutu demonstrated a fair degree of technical knowledge and a high level of competence in cultural

Table 5.4. Knowledge and adoption of certain recommended agricultural practices (percent of all farmers)*

| Technical innovations | | Mazabuka | Akim-Abuakwa | East Kitutu | Wanjare | Teso | Katsina | Geita | Bawku |
|---|---|---|---|---|---|---|---|---|---|
| Manure[a] | K | 100 | — | 100 | 93 | … | 100 | 100 | 100 |
| | P | 92 | — | 92 | 78 | … | 100 | 13 | 75 |
| Fertilizer[b] | K | 88 | — | 97 | 97 | 93 | 100 | 93 | 95 |
| | P | 78 | 2 | 58 | 18 | 0 | 90 | 57 | 7 |
| Seed dressing[c] | K | — | — | — | — | — | 100 | — | 42 |
| | P | — | — | — | — | — | 82 | — | 3 |
| Crop insecticide[d] | K | 40 | — | 70 | — | 85 | — | 100 | — |
| | P | — | — | 32 | — | 48 | — | 12 | — |
| Storage insecticide | K | — | — | 92 | 45 | 5 | 62 | — | 70 |
| | P | — | 75 | 82 | 32 | 0 | 27 | — | 13 |
| New varieties, maize | K | 93 | | 93 | 97 | | 32[e] | 2 | |
| | P | 58 | | 82 | 41 | | 3[e] | 2 | |
| New varieties, other[f] | K | 93[g] | | 82 | 37 | | 77 | | |
| | P | — | | 17 | 17 | | 45 | | |
| Sowing date[h] | K | | | — | — | 100 | | 87 | |
| | P | | | 97 | 58 | 82 | | 87 | |
| Spacing[h] | K | | | 100 | 48 | | | 57 | |
| | P | | | 97 | 22 | | | 5 | |

| Row cultivation[h] | K | | 100 | 97 | 100 | | | |
| | P | 100 | 100 | 87 | 95 | | | |
| Ox plowing | P | — | 35 | 17 | 100 | 50 | 18 | 73 |
| Tractor plowing | P | 33 | 4 | 0 | 23 | 0 | 15 | 5 |

*Data from Food Research Institute field studies. K = Knowledge; P = Practice.

[a]It should be remembered that all cattle-keeping people are familiar with the use of manure as fertilizer for growing crops and most use it to some extent.

[b]Kisii, tea or maize; Teso, Geita, cotton; Mazabuka, cotton or maize; Bawku, Katsina, peanuts.

[c]Peanuts.

[d]Kisii, maize; others, cotton.

[e]Sorghum.

[f]East Kitutu, Katsina, Mazabuka, peanuts; Wanjare, manioc; Akim-Abuakwa, Amazonia and cocoa hybrids.

[g]Aware of variety names.

[h]Kisii, maize; Teso, Geita, cotton.

— Not available.

. . . Negligible.

Blank indicates practice is not among those recommended by extension service.

*Table 5.5.* Extension workers per thousand farmers*

| Extension workers | Mazabuka | Akim-Abuakwa[a] | Kisii | Teso | Katsina[b] | Geita | Bawku |
|---|---|---|---|---|---|---|---|
| Senior | .35 | .12 | .10 | .11 | 0 | .46 | .06 |
| Junior[c] | 5.54 | .12 | 1.44 | .40 | .20 | .64 | .72 |
| Total | 5.89 | .24 | 1.54 | .51 | .20 | 1.10 | .78 |
| Ratio of senior to junior staff | 1:14 | 1:1 | 1:14 | 1:36 | 0:1 | 1:14 | 1:13 |

*Data from Food Research Institute field studies.
[a]These figures exclude the cocoa workers who execute the government plant protection programs. The number of farmers is computed by assuming that 85 percent of the total population is engaged in farming and that the average household size is eight persons. The training of three agricultural assistants and two field assistants is guessed.
[b]One *mallam* per subdistrict. These are agricultural teachers, but without agricultural training. One of the *mallams* was only in charge of soil conservation.
[c]Of the total junior staff of 94 in Mazabuka, 93 were demonstrators who had only attended a specialized course just before the season opened and received some in-service training; 85 of the 144 junior staff in Kisii were similarly "untrained," as were 2 of the 5 in Akim-Abuakwa.

practices which affected their most important crops—coffee for Wanjare and tea and pyrethrum for East Kitutu. Among Wanjare farmers there was an extremely good awareness of maize and coffee practices but a general vagueness about the reasons for coffee spraying, an activity that was carried out by the cooperative on behalf of the farmers.

Knowledge about maize culture was unusually good in both areas. The ready adoption of "maize fertilizer" and "tea fertilizer" was associated with the profitability and availability of these inputs. A short-term credit arrangement with a commercial agent made purchase easier. Coffee farmers generally knew about the recommendations for mulching, pruning, and picking of cherries but tended to ignore them because they were uneconomic in terms of the labor they required.

The adoption level in each of the two sample areas in the district is related to the profitability of each crop grown. Less was known about new varieties of manioc, millet, sorghum, and potatoes than about the more profitable enterprises, including exotic cattle.

*Teso.* The ox plow is used in all land preparation, but the adoption of other ox-drawn equipment, mostly recent introductions, has been limited by availability and technical deficiencies of the equipment.

The number of farmers who apply manure to their crop and the amount of manure applied is small in relation to the number of cattle owned. There are reasons for this: The herding arrangements under which Nyankole herders manage the cattle for their owners ensure that little manure is readily available for the farms, and lack of carts limits the application of manure by those who manage their own cattle.

Awareness of the importance of early sowing of cotton, the right plant population, and thinning is excellent; but the practice is limited by labor constraints, particularly the competition for labor on food crops. Although the ox plow is in general use, there are few planters and weeders.

A generous subsidy on DDT has promoted widespread adoption of the spraying schedule for cotton. Frequent price changes, however, and uncertainty over prices and a natural tendency of farmers to economize in the use of a purchased input combined to reduce the use of DDT below economic levels.

*Katsina.* Ox plows were widely used and their use was spreading. Knowledge of fertilizer use was almost universal, and each season fertilizers were purchased by an increasing number of farmers. Over the ten years 1963–73, sales of superphosphates increased rapidly as a result of government subsidy. This spread of fertilizer use occurred despite a very thin extension coverage. However, supply problems retarded the rate of extension of fertilizer use and of other purchased inputs, including seed dressings and storage insecticides.

The expansion of peanuts, the major cash crop that occupies as much as 30 to 40 percent of the cultivated area, was helped by the availability of the hand-sheller and by a government policy which exaggerated the price differential for premium-grade nuts.

*Geita.* The one innovation that received the greatest amount of attention, propaganda, administrative coercion, and pressure in Geita District after World War II was tie-ridging.[4] Farmers generally understood the reasons for tie-ridging as a result of extension concentration on this innovation for many years, but they did not believe the practice was worth what it cost and they abandoned it when the administrative regulation was relaxed after independence.

Knowledge of recommendations for cotton was poor. Only thirty-four of the sixty farmers knew the recommended spacing on ridges. Although all understood the spraying recommendations to control cotton pests—and eight had spray pumps—they had not been taught how to operate the equipment properly, and only one had sprayed his cotton in the previous season. Fertilizers were issued through the cooperatives, which made their

---

[4]The practice of blocking the furrows between ridges at intervals to reduce runoff and increase percolation.

purchase compulsory by threatening not to issue cottonseed to farmers who did not buy fertilizer.

The method of applying fertilizer was explained to farmers when they received it, but there was no adequate follow-up. Several instances of crop burning by sulphate of ammonia had occurred, presumably because farmers had broadcast it over their cotton. As a result, a majority of the farmers who were interviewed were afraid to use it and were only applying superphosphate. Knowledge of cotton recommendations was poor, knowledge of other crop recommendations virtually nil. This is one area where the lack of extension work was seriously limiting the adoption of farm practices that could increase production.

*Bawku.* The extension service had little to offer to the farmers of Bawku in 1966/67, and between 1962 and 1966 extension was virtually suspended while major efforts were directed to government farms. Field trials and demonstrations, essentially agronomic in nature, were handled by the Crops Research Institute and often supervised from Nyankpala, 150 miles away. The Soil Research Institute was responsible for fertilizer trials, while improved seeds theoretically came from the Seed Multiplication Unit of the Farm Supply Division, Ministry of Agriculture, although there were almost none in 1967.

The existing level of technical knowledge offered the possibility of only very modest increases in farm output or productivity. The high-yielding rice variety D99 was grown generally, and five farmers in the sample of sixty grew high-yielding Spanish peanuts. No improved varieties of millet and sorghum, the staple food crops, were available. Most farmers were aware of the value of fertilizers although only four who lived near the experiment station at Manga had purchased fertilizers. Farmers' awareness of seed dressing was poor. About 70 percent of the sample farmers knew that insecticides could control storage pests, but only 13 percent bought insecticides in 1966. The complaint was that the product was not available. Generally, the awareness and practice of technical recommendations was best in Manga, where gross farm income was highest and employment opportunities were greatest.

## Current Practices

The study areas differ conspicuously in the use of animal or mechanical draft power. Tractors are used to a limited extent in several of the areas, but the important choice is between the hoe or digging stick and some sort of ox-drawn plow. A simple comparison of the importance of this trait with farm income, however, is unrevealing (Chart 5.2F).

Variations in the use of agricultural chemicals—here fertilizers and insecticides—also represent significant differences in technical levels. H. C. Molster has constructed an "agrichemical index" that pools information about the percent of farmers using insecticides and chemical fertilizers (Molster 1969).[5] The correspondence between the agrichemical index and net cash income per farm is very good (Chart 5.2G), especially when account is taken of the high subsidy paid for fertilizer used by Katsina farmers and allowance is made for the large size of farms in Mazabuka and Akim-Abuakwa.

It is not clear which way causation runs in this relationship, although it probably runs both ways. Something can be learned by studying pairs of points (progressive farmers and neighbors). The high rate of adoption of agricultural chemicals in Katsina is the same for both kinds of farmers and seems not to have had much effect on net product. Anthony and Johnston (1968:71) state that "when the rate of acceptance of fertilizer proved slow the timely decision was taken to subsidize the price to farmers and was fully justified by the subsequent rapid increase in demand." It is not clear that use of fertilizers is economic in most parts of the study area. This is a cattle-keeping area with relatively dense human population where manure has been used for generations. Chemical fertilizers may be doing little more than replacing animal fertilizers. Another interesting pair are the progressive and neighbor farmers in Akim-Abuakwa. Fertilizer use is about the same for both, but the difference in income is great.

Perhaps direction of causation can be established by plotting gross cash output per acre against the agricultural index. At the least, this should indicate whether the chemicals are increasing output (Chart 5.2I). The correspondence is about the same as when net income per farm is used as the measure of productivity. The two Katsina groups occupy similar positions, although the relative difference between them is less because it is no longer affected by size of farm. But now Wanjare progressive farmers appear to be in an anomalous position and Wanjare neighbors may be too. The answer probably lies in the fact that the principal cash crop in Wanjare, coffee, has proved not to benefit from fertilizer (nitrogen) applications and

[5]The incidence of use of fertilizers and insecticides was scored from 1 to 5, with 1 representing 0-20 percent of farmers using and 5 representing 80-100 percent. For each group of farmers the score on the fertilizer scale was added to the score on the insecticide scale to obtain the "agrichemical index." This assumes no interaction between the two kinds of chemicals and is not strictly correct. It ranges from a low score of 0 to a high of 10.

Because the appropriateness of the different insecticides—for stored products, seed dressing, or growing crops—varies from area to area, the highest score registered by a farm group for any one of the three uses was taken as the insecticide score (Molster 1969:6).

the return to spraying for coffee berry disease is not high. Uchendu and Anthony (1975b:54) say: "Response to the application of nitrogen is not encouraging, and certainly not economic. Although coffee berry disease remains a major problem, the recommended spraying schedule . . . would cost 265 shillings per acre. . . . Given the kind of yield increases [reported by the Coffee Research Station] one can only conclude that the present level of technical achievements will result in a very small return." Wanjare farmers use few agrichemicals because they have little to gain from them.

For the study areas other than Wanjare and Katsina, the evidence of Chart 5.2I suggests strongly that farmers have higher incomes because they use agrichemicals rather than the converse.

## Access to Markets

Few farmers intentionally produce more than their own families can consume or than they can exchange for other goods. An outlet for farm products is therefore a sine qua non of increased agricultural output, although it may not be of increased agricultural productivity.[6] The distance from the farm to point of sale, the price at which the product can be sold, the quantity that can be sold at one time (both minimum and maximum), and the terms of sale can all be expected to influence the nature and volume of farm output.

The surveys provide some information about product markets in each of the study areas, and additional information is available from other sources. Five of the areas were near railroad lines that were accessible by road. Only Katsina and Bawku were not so favored, but although the Katsina villages are within 75 miles of the railhead at Kano, communication is over a surfaced all-weather road. Bawku, however, is relatively cut off from markets. The railroad terminates 400 miles away at Kumasi, and at the time of the field study the surfaced road from Kumasi to Bolgatanga had been cut by the rising Volta Lake. Truckers had the choice of long waits, sometimes a week or more, for the ferry across the lake, or of traveling the uncompleted alternative road to the north, which was impassable during the rain and destructive of vehicles during the dry season. Bawku and Katsina farmers could sell their produce at local markets, sometimes in their own villages, but the cost of transport to major market centers reduced prices and for Bawku farmers it probably also limited the amount that local markets could absorb. For this reason, cash returns probably represent a

---

[6]Increased productivity may permit resources to be released from agriculture to more highly valued activities while total farm output remains unchanged.

smaller part of the total agricultural production in Bawku than in the other study areas and understate relative Bawku productivity.

Commercial agriculture began in Mazabuka when a railroad was built in 1906 to the lead and zinc mines of Broken Hill (now Kabwe) and on to the Katanga in 1908. This was "a great boon to [Rhodesian] farmers who could, for the first time, rail their maize and beef up to the Katanga mines" (Gann 1964:126–27). Tonga farmers responded to this new market opportunity just as eagerly as did Afrikaners and Englishmen.

Unlike many African areas, the market was nearby rather than overseas and the most important cash crop was a foodstuff. Maize was the "staple food of the African labor force and it was to be the crop through which African producers made a slow entry into the market economy" (Makings 1966:198). By 1925, the amount of maize brought to market by African farmers was substantial and there was a growing use of plows on their farms. As early as 1930, Mazabuka farmers had bought ox carts as well and many were practicing crop rotation. With the collapse of world commodity markets after 1929, a committee of the Agricultural Marketing Board expressed the fear that competition from African farmers would nearly eliminate the European maize grower (Makings:200).

Akim-Abuakwa presents another illustration of the interaction between transportation, effective demand, and agricultural production. This is the area in which Ghana's great cocoa industry was begun in the 1880s. The only communication between this area and the port was by foot, and man was the only beast of burden for cocoa. Later, when the railroad was completed from Accra to Akim-Abuakwa, local farmers decided it was time to construct three large footbridges across the Denus River, a major barrier in transporting cocoa to the railway line. These bridges were not "an old-fashioned form of self-help, they were a newfangled kind of development expenditure, financed by the farmers themselves" (Hill 1956:233). The bridges were built by European engineers and paid for by collecting tolls from "foot-passengers." Two of these bridges were still in use in 1960.

The motor age came to Akim-Abuakwa in 1918, and the Akwapim people are reported to have spent £47,500 on access roads between 1916 and 1926 (Hill 1956:233–34). In both Mazabuka and Akim-Abuakwa the importance of transportation by road or railroad is conspicuous. Akim-Abuakwa records a situation, probably not an unusual one, when farmer-producers recognized the importance of better transport and acted to obtain it without government help.

East Kitutu in Kisii illustrates another effect to be expected from an improved transport system. The initial expansion of tea production in Kenya after World War II had been on estates of 500 to 2,000 acres, and African farmers had been legally prohibited from growing tea (Etherington 1970:6). As the tea estates thrived and output grew, the Department of Agriculture became interested in the possibility of production by small-holders. After considerable experimentation, it was decided to construct factories of minimum economic size to serve small individual tea gardens. One of these, the Nyakoba Tea Factory, is located just outside of Kisii town and it is to this that East Kitutu farmers send their leaves. Tea must be fermented and dried within twenty-four hours of being plucked, and if supplies are to be obtained from many scattered quarter- and half-acre plots in good time and in sufficient volume, a special effort is needed. A network of graded roads was therefore constructed throughout the hills and valleys of East Kitutu along which the tea factory's trucks could travel to gather the day's harvest of leaves. But these tea roads did far more than solve the transport problem for the factory. Roads that can carry tea lorries can carry other kinds of vehicles, and opening up the tea market to East Kitutu opened up the market for maize, pyrethrum, passion fruit, and milk as well. It also made modern purchased inputs more accessible and knowl-edge about new farming methods more widespread.

The use of purchased inputs, including hired labor, in general might be expected to be correlated with and therefore to provide a measure of the development of product and input markets.[7] If this is so, and if develop-ment of input markets stimulates productivity, then the amount of money spent on purchased inputs should be correlated either with gross output per acre or with net cash product per farm. The relationships are shown in Charts 5.2J and 5.2H. There are no surprises in Chart 5.2J, unless it is that the East Kitutu neighbor farmers lie so far above the regression line suggested by the other points. Chart 5.2H, however, shows a very weak correlation except for the effect of farm size on both variables in Mazabuka and Akim-Abuakwa.

## Diffusion of Innovations

Extension services are intended to inform farmers and encourage the adoption of new methods and inputs that will increase net productivity. It is

---

[7]Wages paid hired labor are 70 percent or more of expenditures on purchased inputs of progressive farmers, except in Mazabuka where they are 10 percent and Bawku where they are 13 percent, and 50 percent or more of such expenditures by neighbors, except in Mazabuka where they are 7 percent and Bawku where they are 27 percent. Availability of hired labor might reasonably be expected to increase with expansion of the market economy.

not easy to measure the magnitude of an extension effort because the quality of the staff and the usefulness of its recommendations are at least as important as the number of extension workers. Nevertheless, the number of extension workers per 1,000 farmers provides a crude basis for comparing one area with another (see Table 5.6). The variation is remarkably large but seems unrelated to economic achievement. The willingness of farmers to adopt new inputs depends as much on their availability—probably more— than it does on the propaganda and exhortation of extension officers. The situation in each of the study areas was also very much a consequence of how long farmers had known about the inputs, how long they had been available to them, and how long economically attractive product markets had made use of the new inputs profitable. Some consideration of the history of the introduction of modern inputs into these areas is therefore helpful in understanding the apparent lack of relationship between the amount spent on purchased inputs and net earnings.

## Variation Among Farmers

But farmers themselves differ in their will to achieve higher incomes, their receptivity to new technical methods of production, and their ability to combine the land, labor, tools, and seeds required to grow a crop efficiently. They differ, too, in the skill with which they can identify economic opportunities and in their capacity to take advantage of them. Progressive farmers might differ from their neighbors, then, in their access to productive factors and knowledge and in their freedom to innovate.

Each of the eight field studies reports a wide variation in economic achievement among farmers, whether it is measured in net return per farm or per acre. A great deal of variation is concealed by the general use of averages in the reports. Some idea of its magnitude can be gotten by comparing achievements of progressive farmers and their neighbors, and by examining the variation among other subgroups that is reported in some of the studies.

The general figures for progressive farmers and their neighbors are given in Table 5.3, but the relationships are easier to perceive in Table 5.6, where all figures for gross receipts, expenditures, and net receipts are the ratio of these values for progressive farmers to those for their neighbors. Whether the comparison is made in terms of gross or net cash income, per farm or per acre, progressive farmers in seven of the study areas excel their neighbors by factors of 1.25 to 5.00. Progressive farmers also appear to have more acres under cultivation, although here the ratios are much smaller, from 1.18 to 2.27. Variations of this sort might arise from dif-

Table 5.6. Income and production expenditures of progressive farmers expressed as multiples of income and expenditures of their neighbors*

| | Mazabuka | Akim-Abuakwa[a] | Kisii E. Kitutu | Teso | Katsina | Geita | Kisii Wanjare | Bawku |
|---|---|---|---|---|---|---|---|---|
| Per farm | | | | | | | | |
| Gross cash income | 4.78 | 4.34 | 1.46 | 1.96 | 3.69 | 2.49 | 4.33 | 5.21 |
| Production expenditure | 4.83 | 3.37 | 2.64 | 2.68 | 4.75 | 2.18 | 8.37 | 6.57 |
| Net cash income | 4.74 | 5.00 | 1.09 | 1.81 | 2.90 | 2.55 | 3.63 | 4.96 |
| Per acre | | | | | | | | |
| Gross cash income | 2.29 | 3.70 | .66 | 1.31 | 2.26 | 2.09 | 3.39 | 3.10 |
| Production expenditure | 2.32 | 3.00 | 1.18 | 1.53 | 2.94 | 1.60 | 6.55 | 3.00 |
| Net cash income | 2.27 | 4.20 | .49 | 1.25 | 1.71 | 2.17 | 1.82 | 3.12 |
| Farm size (acres) | 2.67 | 1.18 | 2.22 | 2.70 | 1.77 | 1.18 | 1.29 | 1.72 |

*Data from Table 5.3.
[a]Figures exclude one progressive farmer's income of N₡3,600 from the sale of cabbage.

ferences among farmers in economic opportunity or in economic ability and motivation. It becomes necessary, therefore, to consider whether or not the farmers classified as "progressive" were favored by their society in their access to the factors of production, markets, or knowledge.

The special treatment received by progressive farmers in Mazabuka may partially explain their much greater net output per acre (2.27), but it is not very helpful in explaining the reverse situation in East Kitutu (0.49) or the trivial difference in Teso.[8]

If preference in access to the factors of production is one cause of greater income, it might be expected to show up in farm size, and it does, although weakly in Akim-Abuakwa (1.18), Geita (1.18), and Wanjare (1.29). The text of the studies makes clear, however, that in most of the areas a farmer's access to land is limited only by his labor force (see Chart 5.2A and Table 5.1). Availability of land is probably a serious constraint only in Kisii and possibly in Bawku. In Akim-Abuakwa and Katsina, although population density may be high, there is a developed land market, and the sale of land is increasingly common in Kisii. Progressive farmers probably tend to have more acres than their neighbors because they want them and have the labor to farm them or the money to buy them.[9]

It must be accepted, however, that farms with greater gross value of product enjoy certain advantages in marketing their produce. When a farmer offers more than 150 bags (14 tons) of maize for sale and buys 35 bags of chemical fertilizers—as many in Mazabuka do—he can expect better rates from transporters and better prices from merchants than if he had sold 15 bags of maize or bought 5 bags of fertilizer. These are economic advantages that reflect lower real costs of marketing products and inputs, and they may help to explain why progressive farmers in Mazabuka and Katsina enjoy higher net returns per acre than their neighbors. They do not go very far in explaining the higher returns per acre in Akim-Abuakwa, Geita, and Wanjare and are completely at variance with the situation in East Kitutu (Chart 5.2B)

Evidence that social or political discrimination is an important factor in variation within an area is weak, suggesting that the explanation may lie elsewhere. Some of it must result from intra-area variation in soils, cli-

---

[8]In a study of Rhodesian farmers published in 1968, B. F. Massell and R. W. M. Johnson demonstrated convincingly that the similar distributions between skilled, semiskilled, and unskilled farmers were poorly correlated with farmers' achievements (Massell and Johnson 1968:60–61).

[9]The households of progressive farmers are consistently larger than those of their neighbors. In Mazabuka and Bawku they are about twice as large.

mate, and distance from market. But so far as possible, the samples were drawn so as to have equal proportions of progressive farmers and neighbors in each subarea.

There is little direct evidence that individual differences in the aspirations and skills of farmers are the major determinants of farm returns, but there is some. This explanation is also supported by the subjective appraisals of Anthony, Johnston, and Uchendu.[10]

The variability among farm practices is much greater than is indicated by simply comparing progressive farmers with their neighbors. In the complete Mazabuka sample, for example, maize yields per acre ranged from less than five bags per acre to more than twenty, and cattle herds from less than ten to over seventy (Anthony and Uchendu 1970:254–55). Farm size in Bawku varied from less than six acres (eleven) to more than twenty-one (six). Cocoa groves in Akim-Abuakwa occupied from less than five acres (four) to more than fifty acres (five) per farm and enjoyed yields of as little as twenty loads (of sixty pounds) or less (seventeen) and as much as fifty-one and more (eighteen) and cocoa income ranging from less than N₵100 to more than N₵1,000. In Katsina, thirty plow farmers cropped an average of twenty acres while thirty farmers without plows averaged only seven acres. But one plow farmer cultivated less than five acres and fourteen averaged less than fifteen, while one farmer without a plow cultivated more than fifteen acres and nine cultivated more than ten acres. The distribution of farms by size of farms in Kisii and Teso was recorded and is shown in Table 5.7.

That farmers do vary in their desire for gain and in their technical and managerial skills should not have to be argued, but much of the literature tends to ignore this variation, perhaps because of a preoccupation with averages. There is a widespread belief, too, that incomes of rural African farmers all tend to be the same because family claims compel any individual whose income rises above the average to share the surplus with his relatives. Polly Hill has attacked this "myth of an amorphous peasantry," and her study of Batagarawa village in Katsina provides convincing evidence that there, at least, individual variation in motivation, ability, and achievement is the rule (see Chapter 3, "Productive Activities of Rural Households"). De-emphasis of managerial skills has also been a peculiar consequence of the concern with allocative efficiency and consequent cor-

[10]Their views must be respected despite the bias in favor of this view that most products of Western culture have. That Uchendu's early environment was Igbo and Anthony's Welsh does not weaken this caveat in the least.

*Table 5.7.* Percentages of farms of various sizes in Kisii and Teso

| | | | | Kisii | | | | |
|---|---|---|---|---|---|---|---|---|
| Farms | Under 2.5 | 2.5–4.9 | 5–7.4 | 7.5–9.9 | 10–14.9 | 15–35 | Largest farm | Mean |
| | | | | | (*acres*) | | | |
| Wanjare | | | | | | | | |
| Progressive | 7+ | 20 | 13 | 27 | 7+ | 27 | 25 | 9.9 |
| Neighbors | 4 | 36 | 31 | 9 | 2 | 18 | 35 | 7.7 |
| East Kitutu | | | | | | | | |
| Progressive | 0 | 7 | 33 | 13 | 13 | 33 | 33 | 14 |
| Neighbors | 0 | 38 | 38 | 7 | 13 | 4 | 29 | 6.3 |

| | | | | Teso | | | |
|---|---|---|---|---|---|---|---|
| Farms | Under 5 | 5–10 | 11–15 | 16–20 | 21–30 | Over 30 | Largest farm |
| | | | | | (*acres*) | | |
| Progressive | 0 | 20 | 0 | 7 | 27 | 47 | 80 |
| Neighbors | 13 | 51 | 11 | 7 | 11 | 7 | 50 |

rection for management bias when fitting productive functions. Massell and Johnson (1968:53) point out:

> Efficient allocation on the average farm is a necessary but not a sufficient condition for efficiency on individual farms. For example, some farms may use too much land for one crop and other farms may use too much for another crop, yet the average allocation of land may fully conceal these inefficiencies. . . . There is a clear need to explain the dispersion of farms about the average. When we are able to understand the factors responsible for interfarm differences in resource allocation, then we will know a great deal more about how to increase farm output.

They suggest that T. W. Schultz may have failed to make this distinction when he found David Hopper's study of agriculture in an Indian village supportive of his belief in a "poor but efficient" peasantry (Schultz 1964), for Hopper "examined efficiency only on the average farm. His results are consequently consistent with inefficiency on individual farms."

Kenneth H. Shapiro undertook an investigation of efficiency and modernization in Geita District in 1970/71 based on an intensive year-long study of the activities of seventy-six farm households (Shapiro 1973). Each farmer's technical efficiency was rated on a scale of zero to a maximum of

Table 5.8. Technical efficiency rating
of 37 farmers in Geita District,
1970/71*

| Score | Number of farms |
|---|---|
| 23 to 38 | 3 |
| 39 to 54 | 12 |
| 55 to 69 | 6 |
| 70 to 85 | 6 |
| 86 to 100 | 10 |

*K. A. Shapiro, personal communi-
cation.

100.[11] Scores of the thirty-seven farmers for whom technical efficiency
was calculated are shown in Table 5.8. There is a strong positive correlation between efficiency ratings and modernization scores that measure
social, economic, and technical aspects of farmers' behavior. Shapiro
(1973:412–13) says:

> This does *not* lead to the seemingly obvious conclusion that the compos-
> ite modernization scores directly reflect modern farming and that this
> leads to high technical efficiency.... Technical efficiency in cotton
> farming depends largely on efficient use of traditional inputs and tech-
> niques. Therefore, the correlations... indicate a strong relationship
> between the extent to which a man has moved toward a more modern
> way of life in general and the extent to which he excels in the traditional
> way of farming.

## General Evaluation of the Study Areas

The quantitative information provided in the preceding sections gives
some clues to the differences in performance among areas, but it tells only
part of the story. Product markets are important and agrichemicals contribute effectively to raising net product. African farmers, whether in the forest
of Ghana, the highlands of Kenya, or the miombo bush of Zambia, demonstrate the ability to perceive economic opportunities and to respond to
them. Agricultural growth may be stimulated by well-designed extension
programs as in East Kitutu, by new product markets as in Mazabuka and in
Akim-Abuakwa, or by increased access to one of the factors of production.

---

[11]Technical efficiency was calculated as the ratio of actual output to the output that could
have been produced with the same inputs on the optimal outer-bound Cobb-Douglas production function, constructed according to C. P. Timmer's linear programming methodology (see
Timmer 1970).

In each area, however, the extent of the increases in farm output is closely related to the magnitude of the new opportunities open to farmers.

In order to make possible a more balanced understanding of the information provided by the case studies, the general comments by the field investigators are summarized in this section.

## Mazabuka

Mazabuka farmers have worked beside European farmers for seventy years, and many of them have at one time or another worked for European farmers. The first Europeans to farm in Zambia relied on draft oxen, and Tonga farmers learned the practice from them. Later, when the Europeans adopted tractors and trucks, African farmers bought their used ox-drawn equipment and sometimes their trained oxen, but it was not long before some of them owned tractors and light delivery vans, too. European farmers constituted a large enough market for purchased inputs to support agricultural supply firms, which supplied black and white farmers alike with seeds, chemicals, and farm tools.

In 1967 the Mazabuka District presented a picture of active and comparatively rapid agricultural change that had its roots in the early part of the century. An increased demand for maize, which coincided with the availability of ox-drawn implements, made for an increased acreage of the maize crop. During this early period a familar pattern of agricultural development emerged. Increased agricultural output was achieved by the expansion of acreage and yields remained low.

Alienation of land, natural population increases, and a tendency to move to fertile soils along the rail line and near markets led to localized high population densities. At the same time there was a rapid buildup of Tonga cattle herds, which resulted in overstocking in some areas. This led to a government soil conservation program. The major works put in, and particularly the provision of cattle-watering points, benefited all farmers.

The Tonga are a strongly individualistic people and there are no strong barriers to inhibit personal initiative. Taking the opportunity presented by the availability of a market for maize, a small class of comparatively large farmers had emerged by the lae 1940s, and some were already selling several hundred bags of maize. By 1950 a few had gross incomes of £1,000 or more per annum (Colson 1962:610). Most of this came from the sale of maize, but poultry, cattle, and pigs provided important secondary sources of income. A few farmers had invested in trucks and one or two in tractors, with which they did custom plowing for their neighbors. With the incomes obtained from farming, capital investment in farms continued. This took the form of better housing, wells, the occasional windmill, fencing, implements, and machinery.

In the 1960s, farmers were introduced to major technical innovations, hybrid maize seed, and a new cash crop, cotton. With the use of hybrid varieties and fertilizer, yields of thirty to forty bags (200 pounds) per acre became obtainable. The two inputs, double hybrid seed and fertilizer, cost about £9 per acre at the recommended rate. However, the survey showed that substantial and increasing numbers of Tonga were purchasing both inputs, and that individual farmers were getting yields of over twenty bags (1.8 tons) per acre. The wider use of these innovations was stimulated by the provision of substantial credit and facilities for marketing and the purchase of inputs.

The increase in cotton acreage had been slow, but with the improvement of marketing arrangements and the provision of Solubor for the correction of boron deficiency, future prospects for the crop became brighter. Progress will depend on the individual farmer's assessment of whether or not the crop provides a better return to investment than maize. Tonga farmers are increasingly adopting new practices; they are outstanding in tropical Africa in the extent to which they use a wide range of insecticides for the control of soil, field, and storage pests.

Acceptance of change in Mazabuka District has been stimulated by a combination of factors, planned and unplanned. The area is highly suitable for the growing of maize, for which there was an early and continued demand. The arrival of European settlers made a major contribution toward the development of the area. They showed that farming could be highly profitable and provided an example that was far more effective than any government demonstration could be. Many Tonga obtained seed and ox-drawn implements directly from the large-scale farmers. The existence of a commercial farming community has also provided a stimulus for research. A significant development was the creation of specialist advisory services as part of the research branch. In recent years, shortage of labor at weeding time has prompted requests for information on herbicides. While herbicide investigations would doubtless have been carried out even if there had been no commercial farmers, their readiness to use suitable materials gave the work a new urgency.

Innovations like improved varieties and fertilizers can now be expected to make the greatest impact in this developing agricultural economy. They provide a high return on investment and can be used without any major adjustments in the farming system. If such an innovation is sufficiently easy to understand, so that farmers can learn the techniques from their neighbors, it will be self-propagating and a high concentration of extension workers unnecessary. In contrast, the cotton crop required relatively complicated pest-control measures, and fairly intensive extension support had to be provided.

Agricultural change among the Plateau Tonga over the last sixty years has been impressive, whether measured against the situation existing at the beginning of the century or against the experiences of other African savanna areas, and in recent years the contribution made by agricultural scientists has been of key importance. The Tonga farmer has responded readily to economic incentive. Future development in the area will depend on the availability of markets for Tonga produce. At present the district is essentially a one-crop area, but sales of cattle and milk provide substantial additional income.

A controversial aspect of development in the area is the extent to which assistance should be provided for tractor mechanization on small- and medium-scale farms. A minority of Tonga farms in Mazabuka District are large enough for the economic use of tractors, and there are others that would benefit from the use of seasonal tractor hire for land preparation. However, this is a situation that is best left to the enterprise of the Tonga themselves. Past development suggests that, as the need arises, the wealthier farmers will purchase tractors for private and contract work. Government assistance can most effectively be provided by helping the tractor owners and their employees to obtain adequate training in the maintenance and use of their equipment. Easy credit for the purchase of tractors, or a subsidized rate of tractor hire, is likely to result in inefficient use of expensive equipment and waste of farmers' resources.

## Akim-Abuakwa

Akim-Abuakwa farmers have been a part of the international market economy for a long time. The cocoa industry grew up in response to expanding markets in Europe and North America, with African farmers responsible for all productive activities, including delivery and sale at the ports of embarkation. The principal purchased inputs other than labor are seedlings of improved varieties and insecticides. Farmers are sufficiently aware of their advantages to seek them out at district offices. On the other hand, they know little about the increased returns that might be achieved by applying fertilizer, and little is used, although supplies are available from the Regional Crops Production Office in Koforidua. But supplies of agricultural chemicals are not always adequate, and at the time of Uchendu's study, insecticides were in very short supply.

At the beginning of this century, Akim-Abuakwa had a thin population and much unused forest land that proved to be ideally suited to cocoa. Strong external demand for cocoa made production profitable at a time when demand for palm produce was falling, and led Akwapim farmers who had made money in the rubber and palm-oil trade to purchase forest land from Akim-Abuakwa chiefs on which to establish cocoa plantations.

The immigrant farmers continued to invest in land as long as they had the money and could find forest land to buy. Their organizational innovation, the company system, made it possible to raise sums which proved compellingly attractive to chiefs who had forest land to sell. It also provided an institutional insurance for company members when their land titles were contested.

Improvement in transport and communication accelerated the cocoa expansion. But a particularly important factor in the diffusion of cocoa was the small risk it involved. Its production, at least in its early years, did not involve a radical departure from traditional farming practices and its labor requirements did not compete with those of domestic crops.

Cocoa development in Akim-Abuakwa has completed a full cycle. Production reached a peak in the late 1930s. There followed a period of devastation by disease and pests that caused output to decline throughout the 1940s and 1950s despite efforts at rehabilitation.

Two major problems faced cocoa farmers in Akim-Abuakwa in 1967. The first was what to do with the vast areas of abandoned and derelict cocoa farms that act as reservoirs of pests and diseases and cost the government so much to keep under surveillance. The second was the problem of increasing productivity by the adoption of some of the yield-increasing technical innovations that had become available.

The rehabilitation of abandoned cocoa farms is no longer a technical problem. Scientists at the Cocoa Research Institute, Tafo, have shown that cocoa can be established on secondary forests on land which once carried cocoa that was devastated by swollen shoot, and new cocoa varieties mature earlier and bear more heavily than the traditional ones. Nevertheless, farmers in Akim-Abuakwa, whether they are migrants or Akim, do not consider replanting of old cocoa lands worthwhile without a subsidy, although they continue to plant cocoa on forest lands. Replanting a farm surrounded by pockets of diseased and abandoned trees is much more risky than opening up forest land which will give good return to capital before it is threatened by swollen shoot. The picture was different when government subsidy provided a strong inducement for every farmer to replant, thus reducing the risk of infection spreading to rehabilitated areas.

Ghana's cocoa industry owes more of its increases in output to new plantings than to gains in productivity. The extension of the cocoa frontier from its cradle in the Eastern Region to Brong-Ahafo on the western boundary of Ghana reflects the drive of small farmers to increase their incomes by bringing more land into production.

Increased yields can be achieved with higher-yielding varieties, the right plant population, shade manipulation, regular pod picking and drying, pest- and disease-control measures, and fertilizers. However, these labor-intensive innovations put a strain on the traditional use of labor that works on crop shares and was quite effective in the expansion phase of the industry. A one-third share is no longer an adequate incentive for labor that must perform the more burdensome and complex operations required to make a farm disease and pest free. The emergence of farmers who have acquired larger holdings than they can manage exacerbates the problem.

If the technology of cocoa production is changing, that of food production remains basically static. Very few purchased inputs are used and the standard farm tools are still the hoe and the cutlass. Despite the fact that cocoa farming has resulted in a fixed agriculture, food-crop farming continues to follow a long-fallow rotation even where land is scarce, and only a limited geographical specialization of food-crop farming has been achieved.

Neither lack of effective demand nor of productive capacity limits domestic food output. Farmers in Osino have clearly shown that it is easy to shift from cocoa to commercial food production when cocoa is seriously threatened by swollen shoot. The experiences in Begoro District demonstrate that commercial food production can become a specialized economic activity. And the cabbage farmer in Nankese who serves the Accra and Tema markets illustrates the possibilities that may be exploited in the future.

## Kisii

The Gusii operate a smallholder agricultural system in which techniques of production are labor-intensive and mechanization has made little impact. The first important change in agricultural technology was the introduction of better hoes and superior seeds. Imported English hoes started to replace the traditional hoes in 1911, and in 1914 the Kisumu District Commissioner reported: "There seems to be an ever increasing demand for European hoes. Over 30,000 English hoes have been sold and the demand is rising" (Uchendu and Anthony 1975b:88). Free issues of superior seeds were begun in the 1920s and made for substantial increases in output and productivity. Before the enclosure movement of the 1940s, however, technical and institutional aid to Gusii agriculture was limited. Nevertheless, the Gusii became increasingly involved in the market economy through employment in the mines, settler farms, and tea estates, in the army, and through the expanding internal market for maize. Increasing population found an outlet in the hitherto unused highlands and helped to increase output by bringing new resources into production.

Responses to agricultural innovations in East Kitutu represent a model of cumulative change that has two important aspects: the sequence in which innovations have been adopted, and the characteristics of the individual innovations which tend to make them self-propagating.[12]

The speed with which a profitable innovation is adopted will be much influenced by its compatibility with existing techniques, but this is "partly a technical and partly a cultural attribute" (Jones 1963:392). Compatibility is therefore influenced by the order in which innovations are adopted, since adoption of a prior innovation alters both the technical and cultural attributes of a proffered innovation. Gwyn Jones points out that "if an innovation is recognized by a farmer as a natural extension of his prior experience a quicker adoption may be anticipated," and cites the example of the speed with which hybrid sorghum seed was adopted by Kansas farmers who were accustomed to planting hybrid corn (Jones 1963:392). Just so did Gusii farmers' early success with improved seed conditions influence their reactions to new introductions.

The spectacular agricultural achievements in the Kisii highlands, especially in the last fifteen years, lend further validity to this proposition. It was in 1952 that pyrethrum was introduced into the area. Seven years later, 11,500 farmers were producing $200,000 worth of pyrethrum flower on 5,200 acres of land. Pyrethrum provided the capital which made investment in tea possible, beginning in 1957. The receipts from tea and pyretihrum and from the passion-fruit boom of 1963–65 made subsequent investment in exotic cattle possible by 1963. Farmers who own exotic cattle have been prompted to ensure that enclosures are stockproof and to provide wells as a source of water for their stock, and incidentally for domestic needs. To keep exotic cattle also required more complex management: rotational grazing, regular purchases of feed concentrates, and regular spraying against ticks. An artificial insemination program was also accepted as part of the exotic cattle complex, which implies contact with the veterinarian staff. Each successful innovation has predisposed farmers to

[12]Torsten Hägerstrand's (1968) term as used in his article on "Diffusion of Innovation."

accept others and the sequence in which these innovations have been introduced has been critical.

The Kenya Tea Development Authority provides the grower with a package of services which include extension work, tea road development and maintenance, credit, a market with guaranteed price, and a tea factory. Anthony and Uchendu say that "the tea industry remains one of Kenya's best organized smallholder agricultural industries" (Uchendu and Anthony 1975b:39). Purchased inputs in Kisii are provided by private firms and by the cooperative societies. The arrangements vary with the crops. The tea grower looks to the tea authority for essential inputs and the milk producer to the dairy cooperative union. Seed is supplied entirely by a private commercial firm that also handles seed multiplication. Other firms provide fertilizers, insecticides, and advice about their use (Uchendu and Anthony 1975b:49–50).

Conditions in the Kisii lowlands present a contrast. Although coffee plantings expanded from 370 acres in 1951 to 13,000 acres in 1966, this did not lead to other successful innovations and the general achievement compared very unfavorably with the highlands. There is no technical reason why Wanjare coffee farmers should not enjoy a success similar to that of Akim-Abuakwa cocoa farmers, and it is interesting to note that their gross returns per acre are quite similar. In terms of available alternatives, coffee remains an important cash crop in the lowlands, but poor-quality coffee, corruption at the local cooperative, and high maintenance costs reduce farm receipts and discourage farmers' efforts.

Maize is a major crop in both East Kitutu and Wanjare, with a high income potential because of the new high-yielding varieties. This is not properly reflected in income statistics from the studies because the 1966 main maize crop was almost a total loss.

There was more scope for increasing the "farm stock" in both areas. The availability of exotic cattle was seriously limiting the expansion of the dairy industry and farmers were paying black-market prices for cattle of uncertain pedigree.

## Teso

Cotton was introduced into Teso District early in this century and the first ox-plowing school was opened in 1910. Plows were stocked by local cotton gins and shops, and itinerant plowmen employed by the Native Authority repaired plows and demonstrated their use. Although the Iteso had previously participated little in the money economy, they showed an almost immediate receptivity to cotton when it was introduced, and by

1914–15 they were producing a large portion of the cotton output of the Eastern Provinces. It is not clear, however, how much of this was a voluntary response of profit-minded farmers and how much a response to administrative coercion, the need for tax money, and the discovery that plots sown to cotton provide valuable seedbeds for finger millet.

Cotton output rose steadily from 10,000 bales in 1915 to 31,000 in 1935 and nearly doubled between 1935 and 1965. However, the gains in output per acre were small. The use of manure was popularized, though the extent of its use is still a subject of controversy. Strip cultivation, involving bunding and plowing on the contour, changed the physical layout of the farm, and by 1955 had become general practice.

The Iteso have attracted much attention because of their early receptivity to cotton and the rapid diffusion of the ox plow, which made increased cotton output possible. The introduction of the plow lifted one constraint, that of labor for land preparation, but it created new ones that have yet to be corrected. It helps little to sow a larger area to cotton if this increases the area that must be weeded and picked.

Agricultural development centered on cotton did not lead to the development of other farm enterprises. There are few substantial Teso merchants, and rural petty traders are still a missing link in the growing exchange economy. Limited effective demand at the farm inhibits the exploitation of new maize and sorghum varieties.

Opportunities for investment in the Teso economy are limited. The most attractive investment is in cattle, given the rate of return to crops and the labor demands, but the traditional communal grazing system prevents the adoption of modern husbandry practices. The mixed-farming solution of integrating livestock with arable farming is not feasible because of the large human and animal population relative to available land.

## Katsina

Two features of agricultural change in Northern Katsina are of special interest. One is the emergence of peanuts as an important cash crop without the stimulus of any major technical innovation in cultivation methods.[13] The adoption of peanuts as a cash crop resulted from the growth of a market. More recently, the widespread use of commercial fertilizers has led to greater crop yields.

The farmers of Northern Katsina had adopted a semi-intensive farming

[13]There was a major technical innovation outside of agriculture—the construction of a railroad from Kano to Lagos.

system before the arrival of Europeans. Cultivable land around the villages was cropped almost continuously and its fertility maintained by the use of manure. Fields outside this area and too distant to be manured were periodically fallowed.

The largely self-sufficient nature of these villages, with their multiplicity of trades, creates an impression of extremely conservative communities likely to resist change of any form. On the other hand, economic exchange is highly organized, and both farmers and merchants respond readily to the stimulus of price. The Hausa states have long traded with the outside world, and their merchants have displayed great acumen. An equally striking feature of the society is the importance attached to giving children a Koranic education.

General acceptance of the Muslim faith and the absence of strong lineage ties seem to have been important in the acceptance of change. It has led to a feeling of unity among the people and facilitated communication with other villages and mobility of individuals. Families are not bound to particular localities by the ties of clan land, and this has made it easier for individuals to buy or sell land and to seek economic opportunities in other environments.

Mixed farming was actively promoted in northern Nigeria by the Department of Agriculture beginning in 1928. This was a system in which it was intended that farmers supplement human energy with animal energy through the use of draft oxen, thus increasing the area of land they could cultivate. The animals would be "fueled" with sorghum stalks, peanut and cowpea tops, and bush grazing, supplemented by hay and the tops and tubers of sweet potatoes. Manure would be spread on the fields to maintain soil fertility (Anthony and Johnston 1968:24). The number of "mixed farmers" in northern Nigeria grew slowly to a total of 7,000 in 1950 and about 40,000 in 1966, but farmers who own plows form only a small proportion of the total farming community (Anthony and Johnston 1968:26).

Use of work animals and plows made steady progress in Northern Katsina after World War II, but other ox-drawn implements were rarely used. The response of farmers was limited primarily by the availability of land. Because population density is high, even the introduction of the ox plow creates a potential problem through the displacement of smaller farmers.

Promotion of the use of commercial fertilizers began in 1950 with free distribution in selected districts, and subsidized sales characterized the program in the 1960s. In areas where fertilizer was distributed by the Native Authority there were few problems of supply, but Daura Emirate, where primary reliance was placed on private agents, had serious supply and distribution problems (Anthony and Johnston 1968:29).

Sales of superphosphate increased rapidly in Northern Katsina during the 1960s as a result of the heavily subsidized price, although extension coverage was generally thin. Adoption was dramatic in areas that had been the object of concentrated extension work. Fertilizer was increasingly being used on crop mixtures which do not contain peanuts, and a number of farmers were using it to maintain fertility on bush farms. Acceptance of the idea of fertilizer use may have come with particular ease because the farmers were already accustomed to the use of manure for the maintenance of fertility.

The use of the hand decorticator also expanded rapidly, helped by government policy which exaggerated the price differential for premium-grade nuts.

Progress in Katsina has been limited mainly by a lack of innovations that would permit much increase in per capita incomes. The productivity gains from expanded sale of peanuts may have done little more than permit a much larger population to subsist at per capita real income levels that probably differ little from those that prevailed several decades earlier.

## Geita

The last thirty years have seen the complete transformation of what is now Geita District from tsetse-fly-infested bush and miombo woodland, with a sparse population largely confined to the Lake Victoria littoral, to a major cash-crop and food-producing area. With the removal of tsetse fly and the provision of water supplies for human and livestock needs, immigrants flocked from the crowded areas east of Smith Sound. The population increased about five times from the mid-1930s to the mid-1960s and was settled in individual farmsteads throughout the district. By 1967 Geita was an important cotton-producing area and exported substantial quantities of cassava to the central areas of Sukumaland.

In the period before independence, government preoccupation with soil deterioration in central Sukumaland led to emphasis on soil-conservation measures and the promotion of cassava and sweet potato cultivation as famine reserve crops. Farmers abandoned these imposed soil-conservation practices with independence, but they continued to observe the compulsory cotton uprooting date, although imperfectly.

Efforts to secure adoption of soil conservation measures by farmers have thus made little impact, but the other achievements of the colonial period were considerable, not the least of which was the promotion of new crops.

The ready acceptance of cotton as a cash crop owes much to the work of the plant breeders at Ukiriguru, who produced varieties with resistance to jassid and bacterial blight. This work, important under any circumstances, has special significance at an early stage of development when the small farmer is least able to bear risks.

The agricultural achievements of Geita District in the 1960s were impressive, and cotton production increased by about 60 percent between 1964 and 1967 largely through increased acreage. Some farmers, however, derived most of their cash income from cassava despite the pressure on them to produce more cotton. The trade in cassava was handled by small merchants who appeared to be more efficient than the cooperative society.

Future increases in cotton production will have to be obtained largely through increases in yields per acre. Individual farmers are already short of suitable land, but there is considerable scope for increasing productivity per acre through early sowing, the use of insecticides and fertilizers, and general observance of the closed season. Adoption of higher yielding practices was limited by an inadequate extension staff and unavailability of modern inputs.

The general acceptance of early sowing has been limited by farm labor bottlenecks. Poor feeder roads make it difficult to open cotton markets before the end of the rains, so that farmers who sow early must store their harvested crop until the market opens.

At present, the economy of Geita District is mainly dependent on cotton production, but with cassava locally important as a source of cash income. Geita could emerge as a major food-producing area, but markets are a serious limiting factor to increased production of food crops other than cassava, and local markets can be easily glutted.

Geita in 1967 was ripe for change. The standard of traditional farming was good and Geita farmers constituted a hard-working rural community.

## Bawku

Bawku District offers only limited opportunity for economic growth. Cash cropping is limited to peanuts and rice. Because of the priority given to the compound (food) farm, peanuts are sometimes planted late, resulting in rosette attack and a poor crop. The land suitable to rain-fed rice is small and unevenly distributed among the communities. Lack of hulling facilities seriously limits production. The long dry season finds most able-bodied people with no productive activities unless they leave for the southern towns, mines, and cocoa farms. The major market centers are in the south and transport charges are high, lowering the price of what the farmer sells and raising the price of what he buys. The agricultural productivity of the people of Bawku is severely limited by natural environment, lack of technical knowledge, and geographical isolation, in particular the distance to the southern market. The poverty of the environment is complicated by a dense population.

Land that is poor and scarce, labor that is abundant and seasonally underemployed, a low level of technology, and inadequate and sometimes nonexistent institutional and technical support combine to perpetuate a near-subsistence economy in which the dominant economic motivation is to minimize risk and thereby ensure survival.

Bawku was first visited by Europeans after 1900. The first agricultural officer toured the district in 1930, and extension work began in 1932. A mixed farming campaign was undertaken in the years after World War II. Progress was slow. There were sixty plow farmers in Bawku in 1938, mostly concentrated in one small area (Uchendu and Anthony 1969). Beginning in 1952, however, the Kusasi Agricultural Development Cooperative that had been established by the government became an effective stimulus to the use of plows. It provided good markets for peanuts and rice and sold plows and accessories to local farmers, on credit to cooperative members. The number of ox plows in use rose from 497 in 1952 to 2,645 in 1960, and farmers in the district owned about 60 percent of all plows in northern Ghana. The sale of groundnuts increased threefold, and sales of paddy rice went from 40 bags in 1955 to 1,354 in 1959. There was also the beginning of commercial fertilization.

Ghana's Department of Agriculture was abolished in 1962 and the major agricultural development effort was directed at large state farms. The mixed farming program was dropped, and small farmers like those in Bawku found it extremely difficult to have their ox plows repaired and impossible to replace them. After the coup d'etat of 1965, agricultural extension was restored and the program resumed more or less where it had left off. At the time of Uchendu and Anthony's study, plows and parts were again available but fertilizers, insecticides, and improved seeds were hard to find and the product markets were poorly organized.

Although it is a "poor agriculture," Kusasi farming is "technically relatively efficient," given the present level of knowledge and technology (Uchendu and Anthony 1969:57). The Department of Agriculture has no better alternative to offer, and the traditional Kusasi three-field system has been adopted by the experiment station at Manga.[14] There was no better

---

[14]Compound farms, bush farms, and intermediate farms that are apart from the compound but not in bush.

substitute for the varieties of millet and sorghum that were grown in 1967, though breeding work was continuing. Under near optimum management conditions provided by the station, yields were not particularly high; early millet averaged 450 pounds of grain per acre, late millet about 350 pounds, and groundnuts about 450 pounds per acre.

Mixed farming was promoted in the district beginning in 1938. But only a partial integration of livestock into farming has been achieved; and given the existing farm size, this may be all that can be done. Although one out of every six compounds keeps oxen for plowing, the amount of land available for year-round grazing is practically nil. Enclosure of pasture is impractical since the carrying capacity of the land is so low that the cost of fencing an area large enough for a family herd would be exorbitant. In the circumstances, the present system of hiring fulani herdsmen to pasture the herds away from the farms seems the best solution.

Many of the plows issued under the Kusasi Agricultural Development Cooperative Society Scheme are now either badly in need of repair or require replacement; and for all practical purposes, the corrals have disappeared. Extension staff is thin and relatively immobile, but the most important limitation is that they have very little to offer.

The impression one gets of Bawku farmers is not a general lack of response to change, but rather the lack of opportunities for beneficial change. But some improvements are possible. Bawku has a hard-working population who have recognized needs and are willing to work for them. What is required is a sense of direction, an effective institutional framework, and risk-reducing innovations that promise at least modest increases in productivity.

## With a Broad Brush

Of the eight agricultural economies examined by Anthony, Johnston, and Uchendu, only three—Mazabuka, East Kitutu, and Geita—seemed to be undergoing progressive change in the mid-1960s. Akim-Abuakwa, Teso, Katsina, and Wanjare had enjoyed periods of economic growth earlier, and while Akim-Abuakwa farmers still enjoyed the consequences of this growth, at the time of the studies they seemed to be going nowhere. Bawku had never enjoyed much prosperity in the past, despite the ox-plow adoptions of the 1950s, and its future in 1967 was bleak.

In the preceding pages the course of events in each of these areas has been examined in some detail and the causes of successes and failures explored. Here an attempt will be made to identify only the most general determinants of progress and some of the implications of the studies.

## Successes and Failures

Economic opportunity is a major factor, perhaps *the* major factor, in persuading farmers to adopt new methods and organization of production. The importance of a clearly successful cash crop cannot be exaggerated. Furthermore, success is infectious, and one success is apt to make farmers more receptive to a second innovation. Financial success provides the means to finance technical and managerial innovations and makes farmers more willing to take risks.

In Mazabuka, economic opportunity was created by the private sector, including the construction of a railway to the Katanga. Governmental assistance to Tonga farmers came late, after an earlier period when they were more hindered than helped. The developing European farming sector, however, stimulated government research from which African farmers benefited too, as they did from the farm equipment and farm chemical firms that came into being to supply the Europeans. The development efforts of the 1950s were beneficial to African farmers, although this was more a matter of lifting constraints that had been imposed by the earlier system.[15]

A similar series of events took place in East Kitutu, but it began much later than in Mazabuka and owed much more to the efforts of government. An earlier expansion of maize production in supplying the requirements of employees of government, business, and European farmers produced no further developments. Then, beginning with the government-sponsored introduction of pyrethrum to African farms in 1952 (it had long been grown by white Kenyans), a series of events steadily expanded the Gusii farmers' economic opportunities: markets, roads, processing plants, a good research base, an effective extension program, and a reliable supply of farm inputs. In Mazabuka markets, infrastructure, technical knowledge, and productive inputs had come primarily from private investors; in East Kitutu they came largely from government.

In Geita, the situation differs again. The economic opportunity was provided by government when it opened to settlement land that could be used to grow cotton for export and cassava for domestic sale. But from there on it was essentially a matter of ambitious farmers seizing the newly available resource and putting it to work with almost no change in methods

---

[15]In 1967 it seemed likely that Mazabuka farmers would at last escape the hazards of almost exclusive cash cropping of maize and would add cotton and dairy products. Government price policy and marketing controls in the 1970s, however, appear to have reversed this move toward diversification.

of production and very little help from government. In fact, the independent Sukuma farmer who declined to participate in the Tanzania government's collective schemes was not likely to be considered worthy of
government assistance. Given the quality of agricultural extension in Geita
in 1967, however, it was a question whether farmers were succeeding
despite the neglect of the Department of Agriculture or because of it. There
was no effective sequence of innovations in Geita and now there is not
likely to be one for the independent farmer.

The farmers of Akim-Abuakwa found their economic opportunity in
production of cocoa for the world market, and in the early periods of the
cocoa industry it is possible to detect a series of innovations resembling
those of Mazabuka in consequences although not in form—migration, land
purchase, new organizational structure, privately financed road construction, and new towns to facilitate marketing of cocoa and consumer goods
as well. Prior to World War II, government played a minor role except in
construction of the railroad. Thereafter, government intervened through the
marketing board to reduce cocoa incomes seriously through the 1950s. At
the same time, government-sponsored research found a solution to the
capsid problem and later led to the selection and development of more
productive varieties of cocoa. Research failed completely in the battle
against swollen-shoot disease, although it did develop promising procedures for fertilizer and shade control that an impoverished extension service has yet to carry to farmers.

Progress in Akim-Abuakwa was brought to a stop by swollen-shoot
disease, but it was slowed before this by export taxes and by diversion of a
considerable part of cocoa export proceeds to the marketing boards.

In addition to the loss of income because of payments into the marketing
board's stabilization or development fund, farmers' resources were further
reduced by the compulsory removal of cocoa trees—the "cutting-out campaign" that was ordered by the government. This abruptly terminated
income from cocoa trees that might have borne for many years longer. As a
consequence, farmers found themselves with little alternative to trying to
reestablish cocoa production again where it had flourished long before.
That they did so successfully is a considerable achievement, but it seems
not to have laid foundations for future development.

It is vain to speculate as to how cocoa farmers might have used the
money that was diverted to the marketing board, but it is hard to believe
that they could have invested it less wisely than the Ghanaian government
did. It is at least possible that some farmers might have developed commercial production of new crops in the old cocoa areas if they had had the

resources. There were some indications in the early 1950s that former cocoa farmers might concentrate on producing food crops for the Accra market.

The story of Akim-Abuakwa is still a story of cocoa; it might have been a story of crop diversification to supply the Accra and Tema markets and coastal markets farther west, had Ghana not experienced the extreme economic and political difficulties that have plagued it since the early 1960s.

The farmers of Teso District also enjoyed an early success in developing cash-crop production, but it was primarily government-inspired and never enabled the Iteso to enjoy the kind of incomes achieved by Ghana's cocoa farmers. Both the carrot and the stick were used to persuade farmers to grow cotton for sale. Expansion of the total area under crops was made possible by the adoption of a labor-saving innovation, the use of draft oxen. Cotton was well established in Teso by 1940 and production continued to grow through the 1960s, but little else happened. A suspicion of private enterprise, and of East Indians, on the part of government officials created legal monopolies of trade in cotton by the ginners, and these reduced prices to farmers. The Lint Marketing Board further diverted large revenues from cotton exports into its coffers just as the Cocoa Marketing Board had done in Ghana, and research and extension were unsuccessful in introducing yield-raising practices into Teso farming. For whatever reasons—lack of market development, lack of profitable new production methods, declining internal security, or tenure rules that inhibited growth of animal husbandry—Teso farming remained locked in more or less the same posture from about 1935 onward.

The Katsina story is similar. Early expansion came because of the development of an attractive market for peanuts. Later the production potential was increased by the widespread use of commercial fertilizers—a land-saving innovation. But it is not at all clear why this did not lead to further advances. Peanuts in Nigeria are marketed by a statutory board, but there is no evidence that this placed an undue burden on peanut growers. Millet, sorghum, and cowpeas are widely traded in this market-oriented society, and their trade is not affected by marketing boards or processing monopolies. It may be that the failure lay in the development of new methods of production. Despite the general adoption of commercial fertilizers, their use does not lead to very large increases in yield and in fact is only profitable because it is heavily subsidized.

Private merchants were responsible for early growth in Katsina, but neither the private nor the public sector found ways to continue that growth.

It is hard to find anything encouraging about agricultural prospects in Bawku. High population density, remoteness from markets, marginal moisture and soils leave little room for maneuver. Quite possibly the best solution is that most often adopted by the people of Bawku themselves: migration to southern Ghana where work is to be found on farms, in mines, in industry, and in public works. Large permanent emigration that would greatly reduce the present population of the district might make it possible for those who remain to move ahead. Little else seems possible.

## The Secrets of Success

In the extremely diverse experiences of the eight farming communities that were studied, success, in the sense of increasing productivity, seemed to be associated with the presence of four general conditions and failure with the absence of one or more of them. These four conditions, these secrets of success, have to do with product markets, the resource base, investable funds, and continuity.

*The first secret of success* is to have product markets change in such a way as to increase greatly the profitability of a crop that farmers are already growing (maize, peanuts, cassava) or that they could grow (cocoa, tea, cotton). This change in the economic environment may be autonomous or it may be induced by activities of private or public agencies, or both. It can take at least six different forms:

1. An increase in aggregate demand for a particular product as a consequence of a change in technology, a rise in consumer income, or a change in consumer tastes.
2. A decline in transport costs between the producing country and world markets or within the producing country, achieved either by major changes in transport or technology or by construction of transport facilities.[16]
3. A reduction or removal of political or legal barriers to sale. In an earlier period, this was an important economic consequence of the unification of fragmentary polities into larger units. Later, imports from African colonies were often given preferred treatment in their metropoles. And with the ending of European rule, prohibitions against the production by African farmers of certain export crops were lifted.
4. A decline of production in other areas because resources are being directed elsewhere (cassava in Sukumaland) or because of worsening conditions of production (maize in Zaïre).

[16]It may also result from the unification of fragmentary polities into larger units with accompanying removal of barriers at political boundaries. This kind of change was very important in Africa in an earlier period but did not play a significant role in the eight study areas during the period of study.

5. A rise in costs of substitutes.
6. A decline in marketing costs as a consequence of increased volume, greater competition, or better infrastructure.

*The second secret of success* is that total productive resources be increased so that the potential output of the favored commodity becomes much greater than could be achieved by simply diverting resources to it from other employment. This can be achieved by increasing the existing stock of resources of land and labor or by employing them in different ways.[17] The available stock of land can be increased by capture (Geita, East Kitutu) or by purchase (Akim-Abuakwa). Capture of labor was not a means that was available to twentieth-century African farmers in the areas studied, but wage labor was, although it often required a change in existing institutions (Akim-Abuakwa).

Labor-saving inputs, particularly draft oxen, were a major resource substitute in the early period of agricultural expansion. Other such innovations that employ substitutes for labor include the tractor (Geita, Mazabuka), various sorts of barnyard processing machines like shellers and threshers, and changes in arrangements for household services that free labor from food processing and hauling fuel and water.

Land-saving inputs (substitutes for land) became important later in the form of new cultural practices that were yield-increasing (timing of farm operations, especially planting, plant population, cultivation, and rotation), cultivation of high-yielding crop varieties, and the use of chemical fertilizers and pesticides.

*The third secret of success* is that a substantial part of the profits of farming, especially the profits of innovation, be available to farmers to invest in new productive ventures as they see fit. This increase in investable funds is likely to improve further the availability of land- and labor-saving inputs as larger effective demand expands the capacity of suppliers, just as the increase in output leads to more efficient and capable product markets.

*There is a fourth secret,* too. If the process of growth is to be sustained, the first three conditions must continue to prevail—new market opportunities must be found, more economically efficient ways of using resources must be employed, and the farm enterprise must continue to gener-

[17]In one sense, the natural resource base is determined by the technology, i.e., by the way in which the physical environment is manipulated by man, so that what may be useful in the environment and therefore counted as a part of the resource base in one society may be of no value or even harmful in another.

ate the funds that are needed to take advantage of market opportunities and technical change.[18]

It may at first seem remarkable that in the search for causes of variability in production earlier in this chapter the only clearly demonstrated relationship was between the use of agrichemicals, net return per farm, and gross return per acre. This one finding is, of course, consistent with the second and third conditions (secrets). The fourth secret explains why other relationships postulated in the preceding paragraphs did not show up in the simple comparison of cross-section data. For example, ox plowing was introduced and did increase energy resources in various areas, but only in Mazabuka did market opportunities continue to expand and new resource-enhancing innovations become available. In all areas except Bawku the first steps toward growth came from new market opportunities, not likely to be identifiable in cross-section data. The percentage of farmers growing the newly profitable crops might show this relationship, but if the new opportunity arose some years in the past, as it did in many areas, its effect would have been worked out by the time of the field studies.

Mazabuka farmers have had access to markets and to productive innovations for more than fifty years, and they achieved a steady increase in output long before the government directed serious extension efforts toward them. The farmers of Akim-Abuakwa have been almost entirely on their own except for the government's destruction of trees affected by swollen-shoot disease and for the development of sprays to control capsids. In Teso, once the first reluctance to grow cotton for sale was overcome, farmers willingly adopted ox plowing. But then the supply of proffered new tools and techniques dried up and so did farmer innovation. Geita displays the eager enthusiasm with which farmers who are accustomed to cash-crop farming, as the Sukuma are, will seize upon opportunities created by a great increase in the resource base. Katsina farmers showed a similar response in the enthusiasm with which they purchased commercial fertilizers when government subsidies reduced the price, demonstrating that even socially uneconomic innovations will be adopted readily if they

[18]Uma Lele's emphasis on extension services, manpower training, and provision of improved technology and of farm inputs have to do with productive resources and relate to our second secret; her advocacy of feeder roads and market services primarily concerns product markets and our first secret; and her view of rural development as a "continuous, dynamic process" conforms with our fourth (Lele 1975:189, 191). It should be unnecessary for us to enunciate our third secret, based as it is on the conviction that farmers are rational and economically motivated, but unfortunately the old stereotype of Marshall's savage still persists, sometimes in high places (cf. Jones 1960:108).

are made privately profitable. Wanjare farmers, too, adopted coffee growing with enthusiasm and might have been willing to go on to new achievements had they not been saddled with inefficient and corrupt coffee cooperatives. Even in Bawku there was progress in the form of ox plowing until government reversed itself and put obstacles in the way of its continuation. And East Kitutu presents a story of riding from success to success, kicked off, it must not be forgotten, by a sudden increase in land resources when highland Kisii was opened to settlement.

The prescription for agricultural progress set forth in the preceding paragraphs depends on the amply demonstrated desire and ability of African farmers to respond to economic opportunity. It does not assume that all farmers will benefit at the same time or to the same degree from the general development process. Farmers vary greatly in their managerial, agronomic, and mechanical skills, in their ambition and economic drive, and in their curiosity and venturesomeness. The number of innovators will always be few, and much of the progress of the sector will result from activities of farmers who are striving to emulate these few and to capture for themselves some of the profits of their innovations. It is thus that economies develop.

# 6. Social Determinants of Agricultural Change

Only a part of the disparity in the income of nations can be explained by differences in quantity and quality of the traditional factors of production. Some of the explanation for the observable inequality must be sought in sociocultural factors, the patterned regularities and similarities of sociocultural systems that make behavior predictable.

Since modern development implies deliberate choice—development by design—the social structure or the total system of rules operating in society at any given time is of crucial importance. Herein lies the dilemma which innovating or premodern social structures face. There may be vague awareness that certain technical innovations might result in a net increment of the goods and services demanded by society. But there is neither certainty of the distributive impact of the return from these innovations among the social groups nor any clear knowledge of the long-run impact of such innovations on the rules which regulate the society. What is certain is that despite its net contribution to society's wealth, economic growth disturbs the preexisting relative position of various economic groups, including the powerful. A society that seeks change must have both the strength to initiate it and the security to accept it.

Analysis framed in a structural rather than a cultural context can lead to differing interpretations of the relationship between receptivity to administrative innovation and social structures. L. A. Fallers (1956), in his study of the Basoga of Uganda, a people with centralized political structure and hierarchical social structure, reaches the conclusion that societies with hierarchical, centralized political structures incorporate the Western-type civil service administration with less strain and instability than do societies without such structures.

Field studies in Zambia, however, have led R. J. Apthorpe (1959, 1960) to opposite conclusions. He argues that noncentralized political systems adopt Western-type administration more speedily than the centralized systems because of the greater flexibility of their political structure which predisposes their members to a wider range of new roles. He advances a hypothesis, derived from structural analysis, to the effect that lack of centralized authority is less an impediment to the reception of modern bureaucracy than lack of achieved status ideas and their correlate, an open form of social mobility (Apthorpe 1960:132).

Modern economic growth requires the existence of appropriate technologies that can have a cumulative impact and a social structure that permits the proliferation of innovations. Technology is important in development, but it can only play a permissive role, thus providing a necessary but not a sufficient condition for economic growth to take place. The sociocultural environment which permits technologies to be created and innovated and the human agents who create or accept and apply these technologies to production are the central factors in development. It is not surprising that the literature on innovations has paid more attention to "receiving" societies and to diffusion than to the technologies themselves, even to the extent of ignoring their appropriateness for the societies to which they were transmitted.

Research in the field of diffusion of innovation seems to have come full circle in the last seven decades. Interest in grand, all-embracing theories of cultural development gradually gave way to modest investigation of specific items or traits, but now the attention to economic development of low-income countries seems to have revived interest in the receptivity of whole cultures to technical innovation. A particular target of the "whole culture" innovation studies has been the peasant society. The general impression conveyed by these studies is that peasant social structures create an insurmountable barrier to change. A portrait of what E. M. Rogers (1969) calls the "sub-culture of peasantry," a synthesis of ten negative cultural syndromes, shows how easily the peasant stereotype can get out of hand. The disturbing aspect of this portrait is the obvious fact that no culture or subculture that operates on such an *entirely* negative value system can survive.

A. G. Hopkin's comments on the P. J. Bohannan and George Dalton thesis, that in Africa the market principle acted only peripherally on trade and did not affect production decisions at all, are pertinent.[1] Hopkins says (1973:52):

> Although Bohannan and Dalton claim that their first two categories [reciprocity and redistribution] are applicable to 'traditional' societies in Africa, neither author has made more than brief use of sources which

[1]The Bohannan and Dalton argument is set forth in their Introduction to *Markets in Africa* (1962). Reciprocity, redistribution, and exchange are Karl Polanyi's notion of the main patterns through which an economy acquires unity and stability (Polanyi 1957:250).

historians would regard as necessary to the analysis of the pre-colonial period. . . . The extent to which the market principle failed to mobilise the factors of production fully is better explained in terms of economics (technological limitations and restraints on demand) than in terms of social controls based on anti-capitalist values.

Diffusion studies which emphasize the characteristics of innovations that are compatible with given social structures have added much more to our knowledge of the relationship between social structure and the innovation process (Katz 1963). So also has the growing body of research on the adoption and diffusion of agricultural practices at farm levels (Jones 1967).

Emphasis on the impact of structural and cultural features of African societies on receptivity to change should not be permitted to obscure the fact that adoption of change may itself alter custom, as C. J. Doyle (1974) has pointed out, and change in one part of the sociocultural domain may affect another part through the economic sphere, sometimes resulting in the sort of reinforcing changes that were observed in Mazabuka and Kisii and that have been reported in Serenje (Long 1968). Hopkins also points out that although "African systems of land tenure undoubtedly underwent important change in the twentieth century . . . these were a consequence and not a prerequisite of export growth" (Hopkins 1973:39).

## The Environment of African Farmers

What are the sociocultural determinants of change among African cultivators and farmers? How do Africans respond to change which they perceive to be profitable and in their overall interest? To attempt an answer to these questions, the characteristics of the environment of change in modern Africa must be specified and the bases of traditional African societies and economies and the orientations of African cultures examined. A number of propositions informed by African data but derived from development theories will be advanced.

Whether the primary unit of rural change be the village or a small identifiable social group that forms part of an ethnic group, the unit of development is no longer an autonomous ethnic or local group but the nation-state. The national context in which modern development is cast provides economic actors both opportunities and challenges. It also changes the directives, if not the concerns, of sociological theories of development.

The economic environment resulting from recent political independence in Africa has experienced an expansion of the "policy space," the areas where government intervention in the development process is regarded as legitimate. Independent African governments regard the whole society and economy within the boundary of their sovereign state as a legitimate field of action. This contrasts with the selective intervention policy of recent colonial history.

Postcolonial African society differs from precolonial society and the colonial political structure in ethnic composition, in development orientation, and in the character of economic and political relationship to the rest of the world. The new African governments are still engaged in the process of establishing legitimacy and control and of reconciling the often conflicting goals of political and economic development. In most instances the new states comprise of populations that were politically separate until brought together by the major force of European rule, and conflicts among these historically distinct, even hostile, communities are a major impediment to political stability. It is not surprising, therefore, that political considerations frequently take precedence over economic ones.

The environment created by the new African states can facilitate or inhibit the acceptance of change among African farmers. Traditional African polities which have been incorporated into the new state structure are increasingly losing their autonomy. This is to be expected as development within any national framework inevitably leads to the displacement of local dependence by national dependence; it also means a wider market for all commodities.

With the villages and local communities losing much of their traditional autonomy to central or national institutions, the beginnings of vertical and horizontal integration of the society and the economy become issues in development. Where successful, this process creates the specialization that reinforces development: the incorporation of new economic roles and the transformation of other traditional tasks which lead to widespread commercialization, as well as the acquisition of improved technology and functionally specific and productive skills. This two-way integration process, which promotes specialization among local communities as well as among different industries and sectors, also links their activities nationally. It inevitably undermines the traditional functions and makes it easier to replace them with more efficient alternatives.

An adequate analysis of the response of the African farmer to agricultural innovations must confront the fact that he is no longer operating entirely in his own environment or within the limitations of his traditional culture. He is not only a decision-maker but the subject of decisions made by others. He must contend with the environment of a new state whose institutions and regulatory powers surround, inspire, and often limit his responses as they shape the production possibilities that are available. This does not mean that the farmer's traditional environment is no longer important or has been totally eliminated. It means rather that it is incorrect to regard the farmer's conservatism as the only or even the most important obstacle to development. Change may be frustrated not only by the wrong decisions the farmer makes, but by those decisions that are made for him by others, frequently by national policy makers.

## African Cultural Background

If the traditional African cultures are still viable within the new state into which they have been incorporated, how can we characterize them? What are their assumptions, and are these assumptions consistent with the goals of development in the new state?

Because of the diversity that characterizes African cultures, Africanists (particularly anthropologists) have tended to refrain from making bold generalizations about them. Generalizations are further inhibited by a conscious desire not to add to the graveyard of stereotypes of the African that are so commonplace in Western societies. As Uchendu has pointed out, "Without severe limitations and clearly specified items or traits of comparison, the term 'African culture' as pertaining to a way of life uniformly shared by peoples of Africa becomes meaningless" (Uchendu 1968:227).

Nevertheless, it is possible, as R. A. LeVine has shown, to identify cultural traits that may be said to be distinctively African, although they may be neither limited to Africa nor universal throughout the continent (LeVine 1961:52). One feature of African social organization which requires emphasis, in view of the prevailing models of "primitive" and "peasant" societies that the development literature portrays, is the size of African ethnic groups. Despite the wide differences in population size among them, many African ethnic groups are demographically large and geographically oriented. Ethnic groups with 1 million population or more are common; and populations of over 5 million are not unknown, e.g., the Akan, the Yoruba, the Igbo, and the Hausa in West Africa. These are not the small, isolated communities of stereotyped development literature.

West African ethnic groups also have a tradition of urban and infraurban settlements and large villages, whereas villages of any kind are absent or of comparatively recent development in most of eastern and central Africa. The demographic and settlement patterns in West Africa tend to strengthen corporate interests in land by the perpetuation of a strong lineage system, and this tends to make land reforms difficult, particularly in areas of high population pressure. Within the new nation-state, the demographic features of African ethnic groups influence the politics of development, particularly the allocation of scarce resources in the new state.

In development, ethnicity is a good servant but a bad master. It can make positive contributions in two ways. First, given an appropriate political and institutional framework, ethnic groups can function as units of healthy economic competition in which new economic roles are learned, thus facilitating the exploitation of expanding opportunities. As LeVine (1966a:2) phrases it, "The conspicuous success of one [ethnic] group may generate intense and potentially disruptive competition between ethnic groups over access to opportunities.... On the other hand, ethnically unbalanced achievements may lead the less successful group to attempt changes in their mode of life." Second, the resolution of the conflict between contradictory value systems, particularly in work habits, and between aspirations which members of different ethnic groups bring to a development environment, often tends to result in solutions that initiate development. Development is initiated by perceptible cultural contradictions which can only exist in a flexible multiethnic environment and which enable one group to capture economic benefits that are not yet recognized by other groups. On the other hand, ethnicity hinders economic development whenever it is allowed to be the criterion for the allocation of economic roles, for such an allocation hinders the specialization and free competition that will lead to more efficient production.

Flexibility in social structure is an important feature of African social systems because of its development potential. This flexibility has also resulted in great differences in the economic achievements of various African groups.

African cultures do not fit the stereotype of the conservative, fatalistic peasant who tenaciously clings to a traditional way of life without any hope of achieving improvement for himself or his descendants. LeVine summarizes achievements in African societies succinctly when he writes (LeVine 1966a:3):

> In nearly all traditional African societies with which we are acquainted, the pecuniary motive was well developed, and competition for wealth, prestige, and political power was frequent and intense. Individuals did strive to better their lot and that of their children, and in many parts of the continent they came quickly to see the possibilities for advancement in the trade patterns, schools, and bureaucratic institutions introduced by Europeans. Compared with the folk and peasant peoples in other parts of the world, Africans have been unusually responsive to economic incentives.... Rather than thinking of Africans as tradition-directed people

perpetuating an ancient and stagnant culture, we might more accurately regard them as pragmatic frontiersmen with a persistent history of migration, settlement and resettlement of new lands.

Although Africans, whether in the traditional or modern setting, may be achievement-oriented, they exhibit striking ethnic differences in the status or achievement symbols they consider worth striving for. Ethnic groups that are occupationally oriented, as the Igbo are, measure achievement by individual performance in economic roles. Others, like the Hausa and the Buganda, are politically oriented and tend to achieve high status by demonstrating capability in playing a role in an authoritarian political system. The broader the structure of opportunities which confers social status, the keener the competition to exploit such opportunities and subsequently the more receptive people are to new roles. From this point of view, occupational transformation is the key to transformation of the social structure. One concommitant is the broadening of the opportunity structure so that a greater variety of productive activities and services can be achieved.

Whatever the basis of status achievement in traditional Africa—and both broad and narrow opportunity structures coexist—there is in all African societies a cultural pressure to convert tangible wealth into intangible prestige symbols. This "cultural conversion" may take the form of an elaborate funeral for a deceased relative; taking a costly title (i.e., buying membership in an expensive, exclusive club); or building up of claims against a large number of people—of which plural wives are an example. The conversion of wealth into prestige symbols is not unique to Africa; it can be regarded as a cultural universal. The proportion of the individual's wealth devoted to this activity—while consistent with the values of African prestige economies—may well be so high as to seriously limit capital accumulation and thus inhibit economic development.

African societies vary markedly in their attitude toward work (cf. LeVine 1966a:3–7). Africa has many industrious farmers, craftsmen, and professionals, but not all African societies accept the Tiv and Igbo work ethic that "hard work does not kill a person." Work habits and attitudes in many African societies tend to be more consistent with the view that freedom from work is a prerogative of high status which must be publicly displayed to manifest one's social power. Growing field evidence shows, however, that the demands of a modern economy are eroding the traditional demand for conspicuous leisure.

The profile of African societies and cultures presented above leads to a number of propositions that should guide any model of change in African agricultural development:

1. A correct picture of African societies differs from the stereotypes of peasant social systems in a number of fundamental ways. African societies are set apart by their kinship systems, their access to land and labor, and their rapid social mobility. Their possession of common holistic cultures as opposed to separate status-linked traditions is particularly important (cf. Fallers 1961).[2]

2. The problems of transition in the economic system go beyond the specialization of economic functions and the integration of these functions to create a more productive national economy. This requires the creation of an institutional framework that will orient traditional economies toward the specialization and market interdependence which are necessary for adoption of appropriate technologies.

3. Africa is a continent that has faced radical change in its political system in the last hundred years and whose political system has made radical demands on the traditional economy and social structure in the last three decades.

4. The unevenness of resource endowment and of the incidence and impact of forces inhibiting and facilitating economic change have created uneven development, thus complicating the inherent conflict that accompanies planned, radical change.

## An Approach to Understanding Sociocultural Innovation

The task is to distinguish the conditions under which an agricultural society is better able to change itself so that it will produce an increasing output of those goods and services that are desired by its members.

It is postulated that change is brought about by individual actors who can discover and will apply new methods to achieve new goals and to achieve traditional goals more fully, even at the cost of "constructive destruction" of the old order. Innovation requires perception by the actor that the outcome will be sufficiently desirable to outweigh its costs. It is assumed that farmers are likely to innovate and that economic institutions will be innovated by society when the returns from change are higher than the cost.

An innovation does not occur instantaneously in a vacuum; it requires a hospitable environment and takes time in its production, adaptation, and diffusion. Diffusion may require the acquisition of new skills, tools, and

[2]Robert Bates believes this cultural unity of African societies was a major factor in the drive for independence from colonial rule and that it continues to be an important consideration in the political strength of national governments (Bates 1976).

materials before knowledge of the new technology can be effectively applied. Emphasis on the socioeconomic environment widens the enquiry from the issue of receptivity of farmers to innovations to a broader issue of whether farmers live in an innovative sociocultural system, and if they do, whether they have access to profitable innovations that are suited for their use. The environment for innovation is the sociocultural system with all of its subsystems (political, economic, religious, etc.); it is the set of political, social, and legal ground rules that establishes the basis for production, exchange, and distribution (Davis and North 1971:6–7). It has three inter-related characteristics: (a) it is flexible but logically and functionally inte-grated; (b) it exhibits patterned regularities and similarities that provide a basis for prediction; and (c) each system has unique attributes that modify prediction made solely on the basis of structural similarities. The time required for the diffusion of innovation is influenced by desire, knowledge, and adaptability of the innovation to the attributes of a particular sociocul-tural system. The impact of the new innovation on the social structure may vary from simple adaptation to existing forms (that is, no change in social structure) to total destruction of elements of the system and replacement of them by entirely new forms. It is not possible to predict the long-term impact of an innovation on the social structure, since an innovative society is continually being confronted by changes that can originate from any of its subsystems.

## The Variables

The variables in this view of the sociological determinants of agricultural change are those characteristics of the society *and of the innovations* that influence the ease with which the farmer can incorporate them into his farming system and the benefits to him that will ensue from their adoption. The variables have to do with the nature and goals of the economic and political orders, the technical compatibility of existing and proposed pro-duction methods, and the nature of the system by which knowledge is acquired (the educational system) and of the ways in which status is ac-quired, roles are assigned, and control over property is specified.

These variables can take on values that are favorable or unfavorable to change. They are mutually reinforcing but not necessarily additive, for some can be absolute obstacles to change no matter how favorable the other determinants may be. On the other hand, it is quite possible for incentives and disincentives to be in balance. This appears to be the kind of situation imagined by T. W. Schultz (1964) in his model of a traditional economy, but it is no different in its consequences from one in which a single disincentive outweighs all incentives. Change occurs when the incentive factors are the stronger; then the society is characterized by widespread adoption of new farm inputs, crops, and procedures, and output grows. Economic roles become increasingly specialized, new structural forms and cultural norms are likely to be generated and some traditional roles to be threatened and replaced by new roles. It needs to be remembered, too, that change in the form of adoption of new methods of production can be reversed if the balance of incentives and disincentives that called it forth is upset. This has happened, for example, when the price of an export crop fell and farmers resumed production for own consumption at the cost of production for market.

## Political Environment

While knowledge of farm-level constraints is important, the development environment is determined to a major extent by national goals as they are set and implemented by political leaders and the bureaucracy responsible to them.

The polity, as represented by the new African state, has preempted responsibility for goal commitment and goal attainment for the society. It must therefore be regarded as the substantive location of modernizing efforts, including agricultural development. In this role, the state must provide an appropriate institutional environment and monitor and innovate when necessary the institutional arrangements among economic units. Tanzania's Arusha Declaration is probably the only example of a body of rules of economic action, addressing itself to Africans rather than to outside investors, that attempts fully to meet these requirements and to define the structure within which economic units or actors can cooperate and compete to obtain additional income that is not available outside the structure.

Many of the basic problems facing agricultural development are not within the farmer's power to resolve. David Parkin's statement (1975:3) about eastern and central Africa that rural and urban areas must be regarded as part "of a single field of relations made up of a vast criss-crossing of peoples, ideas, and resources" is equally true of the rest of tropical Africa. The land consolidation in Kenya, for example, could not have been achieved at the farm level; some argue that a disruption as violent as the Mau Mau Revolt was required to make it possible and acceptable. The establishment of agricultural infrastructure—irrigation works, roads and railroads, storage and processing facilities, extension education, research and credit institutions, and a legal framework that encourages and enforces contracts—can normally only be attempted at the national or regional level.

J. D. Montgomery and S. A. Marglin (1965:261) have advanced the hypothesis that guided development of agriculture requires a "political will to develop" which they measure by the amount of public expenditure in the agricultural sector relative to other sectors, the quality of personnel assigned to the agricultural sector and particularly the assignment of senior officers to the rural areas where the farmers live and work, and the attitudes of the bureaucracy toward agriculture and farmers. There is no doubt that the level of public expenditure on the agricultural sector shows government commitment to agriculture. More important, however, is the orientation and stability of agricultural policy and of the senior personnel serving agriculture.

An appropriate administrative structure should make for the maximum possible decentralization of planning and execution, with farmers and local leaders actively involved at the village, district, and regional levels. But decentralization must be carried out in the context of coordination. Horizontal coordination among administrative units must be achieved as well as vertical coordination from village, district, and through regional to national levels. This ensures that the village is allowed to do those things it can do best for itself, leaving other things to larger units.

Economic development in Africa involves problems of social and institutional integration as well as economic. Free physical movement in a wide market of goods, people, and ideas cannot be taken for granted. The environment that makes this possible has to be created by political action. Nor is it sufficient to build a physical infrastructure of ports, roads, and railways; it is equally important to encourage people to learn to trust each other, by extending the national sanctions which will supplement those communal sanctions that are ineffective across an ethnic boundary.

The policy space of the African state has expanded in the last decade of independence, but some of the colonial economic policies have tended to linger on. Fear of dependence on the market, associated with the colonial policy of district and regional self-sufficiency in staple foods, is one such legacy which limits specialization and prevents the formation of the sort of rational and stable economic policy that may be more important than political stability in the early phase of development.

Whatever form the national policy for agriculture may take—whether it calls for political facilitation of economic growth or for a politically directed economic growth—the experiences of agricultural development in tropical Africa in the last seven decades show that a poor agricultural policy is worse than no policy, and that effective demand for agricultural products leads to increased output of both export and food crops.

## Technical Change

The popular conception of science-based technology as problem solver, an instrument for realizing set goals, can be harmful to development strategy. Technical research often solves problems, but it can also create them. Failure to recognize that the problem-solving aspect of technical advance must be balanced against its problem-creating aspect inhibits the monitoring of innovations, frustrates the cumulative impact of innovations, and eventually prevents their integration into the technical inventory of the receiving culture.

When considering the appropriateness of an innovation for a specific situation, it is helpful to distinguish between (1) technology-centered adaptations, which require that the operations of the farm be modified to harmonize with the available technology, and (2) farm-centered adaptations, which are designed to meet the problems of a particular farming system. The characteristics of suitable technical innovations vary from one farming system to another. In the economies of tropical Africa, where capital is short relative to labor but land is still fairly plentiful, economic specialization has only begun, and opportunities for urban employment are severely limited, the techniques that are most attractive for small farms have the following characteristics:

1. They are highly divisible and neutral to scale and so can be easily adapted to small farms.
2. They are cumulative, so that the adoption of one is eased by the prior adoption of another, thus permitting efficient sequences of innovation.
3. They fit easily into labor-intensive systems and their use can be learned fairly quickly.
4. The inputs they require can be manufactured locally.
5. It is easy to determine how effectively the tools and implements they require are working and to repair or replace them.

The savanna regions of Africa offer great opportunity for the application of intermediate agricultural technologies. Successful efforts have been made to introduce the ox plow in many parts of tropical Africa, where an integrated ox-plow technology for land preparation, weeding, harrowing, and other activities offers great advantages. (The lack of emphasis given to the windmill in this region is difficult to understand.) New methods that one farmer can easily teach another and that employ highly divisible inputs, like improved seed varieties, fertilizers, insecticides, and pesticides, have tended to win farmers' acceptance. An important feature of these innovations is that they reduce risks as well as increase yields.

The traditional means of increasing agricultural output in most parts of Africa is by exploiting the most abundant factor—land. Many communities have reached the limits of their land frontier. But there are vacant lands that could be opened by employing new technologies; three such areas are the tsetse-fly zone, the swamps and riverine areas, and the evergreen forest. The economic value of the physical environment depends primarily on man's ability to use it. In this sense, a country's natural resources are sharply conditioned by its technology. The colonization of Geita District of Tanzania by the Sukuma, following the control of tsetse fly in this area, illustrates what potential for agricultural growth there is in many parts of Africa.

Among the most successful innovations in African agriculture are new crops, especially those of the Americas, whether grown for export or local food supply.[3] Jones (1957) has called attention to the sociocultural factors facilitating the diffusion of manioc and certain parts of the manioc complex in tropical Africa. It is also important to recognize how such new crops as manioc and maize have altered the economic and dietary bases and even the value systems of the accepting societies. The Igbo of Nigeria are probably without parallel in Africa in the way yam cultivation organizes their value system, providing them with a basis for economic competition and allocation of prestige (Uchendu 1965). The yam crop is so institutionalized among the Igbo that it is one of the important foci for "title taking"—an institution that enables the Igbo to convert wealth into prestige. Today the economic advantages of manioc are so overwhelming that the crop now constitutes a serious threat to Igbo yam culture. The economic facts are that the yam is a most labor-intensive crop whose labor demands are no longer consistent with the transitional Igbo economy and society, in which cheap labor has become relatively scarce because of the growth in alternative employment activities for adults and because most children are in school. In many parts of eastern and central Africa where sorghum and millet were the traditional and culturally preferred staples, the economic advantages of maize have helped to alter the economic as well as the dietary bases of these societies. These examples suggest that even in

[3]New World crops introduced into tropical Africa after 1500 include maize, manioc, potatoes, sweet potatoes, new cocoyam (*Xanthosoma*), peanuts, kidney or haricot beans (*Phaseolus* spp.), pineapple, avocado, papaya, and cocoa.

such highly resistant cultural areas as food habits, economic factors can effect a major transformation.

## Relative Supplies of Capital and Labor

Supplies of capital and labor are interrelated and can be substituted for each other. Much of agricultural capital throughout history was produced by farmers and farm laborers working with simple hand tools—sometimes with animal draft power—to clear, level, and drain the land, to construct diversion and storage dams for irrigation water and dig canals to distribute it, to plant groves of economic trees, and to build roads, houses, and other farm structures. Pyrethrum farmers in Kenya and cocoa farmers in Ghana and Nigeria, to cite three examples, were able to generate their own farm capital by the deployment of their labor, including that of the members of their family, on their fields. Farmers of eastern Ghana accumulated investment capital from food-crop farming, which they later invested in the palm produce trade. It was this capital and the accumulated earnings from it that financed the purchase of idle land for cocoa plantings and led to the development of the largest smallholder cocoa-farm enterprise in the world. The widespread use of ox-plow technology in Teso District of Uganda, the successful introduction of smallholder tea enterprises in Kisii and other districts in Kenya, and the growth of rice production in Abakiliki Division of southeastern Nigeria after World War II are select examples of widespread economic activities involving substantial capital investment by the smallholder farmer that received little or no financial support from any government.

The crucial question is not whether the smallholder farmer can accumulate investable funds, but under what conditions he is likely to make a substantial economic investment that implies foregoing various social commitments. The appropriate condition is that the innovation be profitable. It is profitability that led Krobo farmers in Ghana to move from one enterprise to another; that led migrant farmers in eastern Nigeria, whose traditional occupation is yam farming, to become successful rice farmers in Abakiliki; and that led the Kisii of Kenya to adopt successfully a range of agricultural enterprises that have falsified the earlier notion that tea and pyrethrum are "plantation" enterprises and that exotic cattle could not be successfully reared by African smallholders.

The smallholder farmer may not have all the funds he needs. But it is use, not availability, of capital that matters, and most often what inhibits capital formation (which frequently has a large embedded labor component) is the level of profitability. As a farmer's enterprises outstrip his managerial capability, lack of knowledge of how to apply additional capital effectively becomes an important limiting factor.

## Ownership Concepts—Land Tenure

The earliest focus of inquiry on African tenure systems was to determine whether or not Africans "sold" their land. That question proved rather unproductive and was eventually abandoned. Anthropologists still worry about rights in land as a dimension of social structure, but the economist's concern is with barriers posed by land tenure arrangements to the process of economic development. In the latter frame of reference, attention is directed to political, economic, and institutional factors which tend to keep rights in land outside the marketplace or render them immobile.

If there is any single attribute of land tenure that facilitates agricultural development, that attribute can be conveyed in one word: mobility. The question at the early phase of agricultural development is not whether farmers enjoy freehold tenure or have registered titles, but rather whether their tenure arrangements are flexible enough to accommodate innovations. A. G. Hopkins (1973, p. 39) concluded from his examination of the effect of land tenure rules on economic development in West Africa that "African systems of land tenure undoubtedly underwent important changes in the twentieth century, but these were a consequence and not a prerequisite of export growth." A flexible tenure system requires security of interest in land, under terms known and agreed upon in advance, and easy accessibility of rights. In most traditional African economies, the security of rights in land is guaranteed and protected by the very principle under which the initial rights were acquired. In one community it might be the kinship principle, in others it might be the principles of residence, clientship, service to a higher authority, or mere political affiliation or allegiance. As long as the social relations which give rights in land are maintained, the question of insecurity in land seldom becomes a live issue.

Land tenure is a series of relationships between individuals with regard to the ways in which land may be used. Property in land consists of a "bundle" of rights with regard to the same piece of land that can be enjoyed by different right-holders, whether individuals or social groups. African economies have tended to overexploit this attribute of land in a way that poses problems for increased adoption of technical innovations. A wide diffusion of rights in land in many African communities makes certain technical innovations impossible, although it has the advantage of preventing the emergence of a landless class.

Uchendu has argued that technical change tends to simplify tenure arrangements by progressive reduction of the number of right-holders in a given piece of farm land, particularly as change leads to increasing production for sale (Uchendu 1970b). The case studies carried out in connection with this study support this hypothesis, and the agricultural development in Kisii District of Kenya presents an illustration of the simplifying of tenure arrangements in response to technical innovations. The introduction of profitable and high-value cash crops into the Kisii farming system—pyrethrum in 1952, tea in 1957, exotic cattle and synthetic maize in 1963, and hybrid maize in 1964—kept interests in individual farm holdings quite exclusive.

African systems of land tenure differ from classical peasant tenure systems in *structure* and in *content*. Structurally, the land tenure systems of tropical Africa (with Buganda and Ethiopia as exceptions) have no exploiting landlords or helpless tenants. In regard to the content of tenure, multiple kinds of land rights and ways of acquiring them create multiple interests in land. These two features provide adequate opportunity for agricultural innovations to be initiated. The challenge is not to the ability of African land tenure to respond to simple agricultural innovations but rather to its capacity for continuous innovation and adaptation.

It is an easy error to assume that the land tenure system can continue to adapt to new demands without changes in institutional arrangements. Instruments to facilitate the acquisition of rights in land across ethnic boundaries are inefficient and often absent, and local sanctions which protect acquired rights are ineffective. The problem is not always one of crossing an ethnic barrier; many local sovereignties over land may exist within one ethnic group. In this way, the rigidities which hinder mobility of interests in land become a national problem. The concept of local sovereignty over land also makes it possible to appreciate one of the reasons why land-deficit and land-surplus economies coexist side by side in many African states. The challenge to policy is clear: Local sovereignties over land that inhibit mobility of interests in land should be removed by political action.

The African system of inheritance can also pose problems when land pressure increases. The principle that every male child must be ensured of a piece of land is an enlightened social policy which does not depress production as long as there is land for expansion. The application of this principle in many parts of Africa, particularly among the Igbo, the Kikuyu, the Chagga, and the Chigga, to mention a few areas where population pressure is unusually high, has resulted in a fragmentation[4] of holdings that reduces

[4]Not to be confused with the practice, sometimes called parceling, of building up a holding from small plots of different types, e.g., bottom land, hillside land, plateau land, selected on

the economic variability of a holding and poses serious problems for production.

## Demographic Pressure

Rising urban population is a major concern in most African countries, although only a few countries are beginning to recognize the problem that rising national growth rates impose on their economic performance. A half-century ago the problem was different: the issue then was how to get African labor committed to work for wages or money income. Today, it is with the unemployed urban population and with how "to get them back to the villages." But the rapid population growth of recent decades has created population pressure in some areas which has led to rural-to-rural migration as well as an outflow to urban areas. (See, for example, Lynam, 1977.)

Migration to the towns from the countryside is politically visible, besides imposing a strain on urban resources. But migration within the rural sector, which makes scarcely any demand on public institutional resources and often contributes greatly to agricultural output, is often ignored. Although in the past such migration within the rural sector has often led to an increase in per capita as well as total output, there are currently instances where the agricultural potential of the areas of in-migration is so poor that the migrants lead a precarious existence.

Under what conditions does labor migration make a positive contribution to agricultural development? A. K. Mabogunje (1972) argues that the net result of labor migration in West Africa has been positive and that the migrant is a major factor in economic development. We can bring a wider perspective to the relationship between migration and agricultural development if we separate three interrelated issues: the economic impact of demographic pressure, the attitude of the urban population toward farmers, and the skills and capital funds that returning farm laborers invest in their own agricultural operations.

In an early phase, agriculture benefits from population growth. Certain essential activities, such as protecting a growing crop from birds and wild animals, are more easily performed by large communities than by a few isolated families. An increase in numbers also makes possible greater specialization in production of goods and services, with accompaning increase in quality and lowering of costs. And some agricultural systems that require such modifications of the environment as terracing, drainage,

the basis of suitability for particular kinds of farming, scarcity (cost), and distance from the farmstead.

and irrigation are only possible when the communal work force is large. Economies are to be found, too, in concentrations of farming populations that permit large-scale specialization in production and lead to reductions in transport and marketing costs.

Inadequate population is a very real consideration in sub-Saharan Africa, where many areas have too few people to provide an economical market or to staff an economically productive enterprise.[5] Accidents of the timing of colonial rule sometimes caused populations to be peculiarly out of balance with resources, although population density in about 1950 did crudely reflect some features of the environment (cf. Bennett 1962: 211). Prior to European intervention, the people of Africa were in a state of continual action and reaction, to their neighbors and to a technical and economic environment that was changing rapidly as contact with Western Europe became more frequent and more intensive (Jones 1965:104–10). But with the imposition of European rule, many groups were frozen in place, even though they might have been on the way from serious disequilibrium with their environment to reasonable equilibrium. One consequence of this enforced mismatch between people and environment was a major development of labor migration.

At the same time, groups faced for one reason or another with rising population on a restricted land area developed a variety of ways to increase returns to land—now become the scarce factor—and still maintain reasonable levels of living for the community. Response to demographic pressure takes a number of forms—the reduction of the fallow cycle; intensification of cultivation, including the use of manures; production of higher valued crops; and specialized production for market. The Kofyar and Kano farmers of northern Nigeria and the Chigga, Chagga, and Ukerewe of eastern Africa provide select examples of African peoples who have innovated intensive, sometimes terraced, farming systems in response to demographic pressure. These communities are reaching the limit of their ability to modify and improve their farming system with the existing technology, and the only options remaining are institutional and technological innovations. The experiences in Kisii District of Kenya, Mazabuka District in Zambia, and the response of the elite in Uganda, Zambia, Kenya, Nigeria, Liberia, and Ghana show that funds can be obtained to finance these innovations—sometimes from the townsmen—if the potential returns are attractive enough. Successful agricultural innovations in

---

[5]See, for example, Ester Boserup's Chapter 8, "The Vicious Circle of Sparse Population and Primitive Techniques" (Boserup 1965:70–76).

Kisii and Mazabuka districts owe much to the zeal, skills, and capital funds of migratory workers who have returned and invested their resources in agriculture, and the economic success of the returning migrant has often provided a role model that is copied by members of his village.

## Role Determination

The specialization of economic roles is both a necessity and a consequence of economic development. Factors that limit role specialization also inhibit agricultural development. They include social structure, particularly as it caused roles to be determined by ethnicity, age, and sex; the ideology of local self-sufficiency; narrow opportunity structures; and individual or group self-definition.

The model of social structure in developing countries that is presented by the literature on development resembles a three-storied house with a small elite at the top, a few middle-class business and professional men and bureaucrats in between, and masses of ignorant and exploited peasants at the bottom. This model does not portray African social reality. Characterizing African societies as having little social stratification also misses the point, for inequality has always existed in African societies, though only a few of them have made inequality the organizing principle of their social structure.

The tendency to exploit ethnic or social advantage is not unique to Africa. It flourishes in all political systems where "ethnic privilege" is profitable. Anything that lessens the temptation to seek profit by manipulating ethnic links tends to redirect the energies and resources of ethnic groups toward economic competition that is favorable to development (Keyfits 1959: 44).

Economic role specialization is also impeded by continuation of traditional forms of organization that tend to perpetuate old practices. Traditional work groups, for example, are ill adapted to the utilization of new skills. Sowing seed in rows at specified intervals, thinning seedlings, and applying fertilizers may seem simple tasks, but they are functionally specific ones which the work group has neither the incentive nor the training or experience to perform. The use of household labor in some of those tasks poses similar problems.

Management problems also stem from poor organization or from the structure of the farming household, and the managerial ability of many African farmers has not kept pace with the increasing complexity of their farm enterprises. Poor management limits the gains from technical and economic innovations.

A major problem in African social structures lies not in their inflexibility but in their narrowness in terms of the total context of the new state. The African state incorporates various social systems that originally were almost autonomous. The task of economic development is therefore one of social engineering to give these different structures a national focus.

Traditional African societies vary in their opportunity structures. Those that are occupationally oriented tend to adapt to a wider range of economic roles, without any fear of cultural penalty, than those that manifest narrow opportunity structures for status achievement. Sex-linked roles pose problems in the transformation, just as do ethnic-linked roles. However, a change in environment, an ecological disaster, or a powerful economic incentive can alter the self-definition of the individual or groups in a way favorable to development. It is well known that migrants accept occupations and work roles which they would consider beneath their dignity in their home areas and that the destruction of a people's economic base often forces them to make radical adjustments. The "agricultural Masai" provide an example (although not a prescription).

Economic factors that help transform traditional occupations also lead to increasing specialization of economic roles. One of the important changes in African economies in this century is the widening of occupational structures, which also implies the obsolescence of a number of traditional occupations. New crops that have been adopted into the farming system have often led to modifications of the traditional definitions of "male" and "female" crops when their economic potential was large. Cultivation of food crops is no longer exclusively a womens' activity in those parts of Africa where the domestic demand for staple food has led to increasing commercialization. There has also been deliberate specialization in the growing of crops in certain areas that goes beyond the traditional concept of cash cropping. Intensive, specialized farming of single crops which uses purchased inputs and employs technologies that permit all-year crop activities is emerging. Examples include the Anlo shallot farming in southeastern Ghana, rice farming in Abakiliki in eastern Nigeria, and vegetable farming by the Kigezi of southwest Uganda. Stimulated by demand from Europe, export crops have become a major engine of growth, driven by the small African producer wherever he has been allowed to produce these crops. Domestic demand for food crops will increasingly have similar consequences.

## Value Systems and Status

African value systems vary enormously, but none poses insurmountable obstacles to profitable technical change. The major obstacle has been the lack of profitable innovations that can support change and thus expand the value system. It is expanding opportunity that transforms the value system, not vice versa. Values change because of better alternatives and wider choice.

It is often asserted that Africans fear and resent the wealthy, and that there is social pressure to share wealth among a wide retinue of kinsmen, thus limiting the individual's desire to accumulate wealth. Another version of this "fear of the wealthy" complex is that successful innovators tend to become targets of aggression by the poor. The weakness of these hypotheses lies not in the ethnographic observation but in the assumption that it prevents the accumulation of wealth. If these behaviors are so important, why is there such a wide inequality in wealth in African societies—both traditional and modern?

Social change does not occur without conflict. Rapid social change upsets the rules of the game in an effort to establish new rules of competition. It is for this reason that those who stand to gain from social change demand it, and those who tend to lose, oppose it. In Teso District of Uganda, enclosure is resented by those who feel that the wealthy are fencing them out of their traditional rights of free access to grazing land. In this society where the rules of tenure have not changed, even though the economic environment has been altered, it is understandable that some Teso farmers resent enclosures. This is not the same as saying that the wealthy are targets of aggression because of their wealth. D. R. Smock observes that among the people of eastern Nigeria "there is enormous desire to accumulate wealth and little hesitation to display it" (Smock 1969:117). In his study of staple food crop marketing, Jones did not find the charge of extended family parasitism supported. He writes: "The important fact is that a great many Africans are able to carry on successful trading businesses, however persistent the claims of kinsmen may be. Furthermore, there is general evidence of family cooperation in trade whether in the form of advancing funds or in the acceptance of kinsmen as partners, apprentices, and employees" (Jones 1972:239). The Gusii of Kenya provide another example of an African society where rapid rural change has been achieved by a people whose culture emphasizes ancestor worship, authoritarianism, interpersonal hostility, pervasive witchcraft, attribution, and jealousy of the wealthy. As R. A. and B. B. LeVine report, the Gusii fear of the wealthy "does not inspire the wealthy to share with others less fortunate [but] on the contrary they use their wealth to dominate their inferiors through loans and threats of expensive litigation" (LeVine 1966b:11).

What is the impact of African kinship and the extended family systems on economic development? The general conclusion reached by those who use the social system model is that the extended family is dysfunctional to development, and that this type of social structure fosters parasitism and discourages social mobility.

Uchendu (1970a) reaches an entirely opposite conclusion. He maintains that it is the African elite themselves who, having achieved their present position through the sacrifices of extended family members and exploitation of the cheap labor this system provides, are principally responsible for the view that the elite are being preyed upon by members of their extended families.

Many investment decisions in tropical African agriculture, particularly in the area of adopting permanent cash crops, are not made completely selfishly. Elderly men and women who had no formal education have contributed financially to the construction of village schools for their children; in a similar way they have provided for the future by planting cash crops that are unlikely to come into bearing until after their deaths. The overriding motivation in these decisions is a sense of family commitment, a desire to maintain the continuity and viability of the lineage; they are motivated by the desire to leave something for their children. Further, in those West African cultures that subscribe to belief in reincarnation, individuals are encouraged to leave to their family a place that is better than the one they found, since they themselves expect to return to it.

The extended family is not only the source of major farm inputs like land, capital, and education; it also provides constant reinforcement that helps in goal attainment. It sets new goals and aspirations for its members, and, as LeVine (1966a) points out for Nigeria, the extended family is an important, if not an indispensable, agent in achievement motivation.

Rising productivity and output may be goals of an economy, but they are not adequate incentives for the farmer. He is interested in producing more cocoa, coffee, or cotton because of what it brings to him—money and enhanced respect. If the farmer's interest in production is to be sustained, the opportunity for him to demonstrate success in either the traditional or the modern style—and preferably in both—should expand as his productivity and output expand. The history of agricultural development in Teso District of Uganda suggests that the satisfaction of traditional goals alone does not provide sufficient stimulus for continuous expansion of output. The Teso showed remarkable receptivity to the early innovation of ox-plow technology, which enabled them to produce more on more acres. New to the money economy, the Teso invested their increasing earnings from cotton in one traditional asset that afforded them the greatest incentive to work: cattle. But there are few things once can do with cattle in Teso-land, and the cultural preference called for investment in marriages. In about three decades following the successful introduction of cotton, Teso society faced an unprecedented inflation—many cattle chasing a constant supply of women. Their appeal to the colonial authorities resulted in bridewealth limitation that everybody ignored. The net result was a disincentive to labor that affected the output of cotton in the 1930s and 1940s.

## Summary

Institutional changes and new technologies which serve a productive agriculture will be devised and farmers will adopt them as long as the economic and sociocultural incentives are, on balance, stronger than the disincentives. The forces that influence adoption fall into three broad categories: (a) those that affect the policy environment in ways that maximize or penalize effort; (b) those that condition the attitudes of the farmer in ways that encourage or discourage his active search for and widespread adoption of innovations; and (c) those that affect the public institutions that generate new methods and materials designed to increase agricultural output.

The productivity and output of an agricultural economy can follow three possible trends, depending on the interaction between incentives and disincentives. First, productivity and output of goods and services may remain unchanged for a long period, indicating a balance of the incentive and disincentive groups of factors. Second, productivity and output may fall below an existing equilibrium level if the existing system can no longer adjust to a changing socioeconomic environment and if disincentives to innovation are stronger than incentives. Third, productivity and output may show an upward trend when incentives are stronger than disincentives. In this interacting system, the farmer is the driver of the engine of agricultural development; if he is to proceed, he requires innovations that are adapted to the physical environment, that are centered on the existing farming system, and that are profitable. Agricultural development is a process of overcoming production constraints wherever and however they may originate. The major ingredients of a progressive agricultural system are the ability of farmers to recognize production bottlenecks, the capacity of the economy to generate appropriate technical solutions of those bottlenecks that cannot be opened by farmers working alone, and the recognition by policy makers that new technologies applied to production must be monitored to prepare for new bottlenecks as old ones are overcome.

One of the important features of the African development experience in this century is its *uneveness*. Geography, culture, and historical accident are no doubt involved in any satisfactory explanation of this phenomenon, but one simple factor which seems overriding is the inequality of economic opportunities. It is much more important than any observable difference in the aspirations of African people.

The seven case studies conducted by Anthony, Johnston, and Uchendu confirm that agricultural development is determined by the net balance between the sociocultural costs and benefits of innovation. The evidence is overwhelming that cash crops have been adopted when they were profitable; that increased output tended to depend on the exploitation of factors of low opportunity cost, particularly land; that sustained demand was necessary for increased efforts and acceptance of profitable innovations; that profitable investment opportunities, whether in land, labor, or in high-value crops and livestock, have been financed without much assistance from government or other institutional lenders; and that in general economic changes have been accompanied by changes in the sociocultural structure.

The level of an economy's output of goods and services depends on the "mix" between two interacting groups of factors: those that are favorable and those that are unfavorable to change. While an economy will alter its institutions and individual farmers will adopt innovations that are viable and profitable, the sociocultural system aspires to achieve more than economic goals. Because of this, profitable opportunities may not be perceived in time; and even when they are perceived, immediate institutional response may be frustrated by culture lag or by vested interest which stands to lose in the inevitable readjustment of social and economic values and resources that any technical change brings about. Understanding of the development process is heightened by moving away from "preconditions" and "prerequisites" to development, and by emphasizing the instrumental role of technology and the view of technology as problem-solver and problem-creator.

# 7. Agricultural Extension Services

In an important sense, agricultural development is a major learning experience. Factors in the sociocultural environment that improve the farmer's knowledge and technical competence include experiences as migrant laborers, farmer education programs that teach specialized functional skills, the methods employed by more successful local farmers who serve as models for the community, and the opportunity provided by the sociocultural system for the demonstration of achievement and prestige in both traditional and modern forms.

African farmers who are able to recognize that their traditional farming techniques and present level of knowledge impose a limitation on their farm enterprises seem to be on the road to success. Successful farmers in Kisii and Mazabuka districts often cited lack of technical knowledge as one of the limitations affecting their agricultural output and productivity.

One of the cultural pitfalls in extension education in tropical Africa is the false assumption that the "farmer" and the male " head of the household" are always the same. Making the male head the target for all extension education and training ignores the fact that on many African farms, women are responsible for sowing the seed, weeding the growing crop, and storing the harvest. It is not enough to tell the man that planting crops in rows and early weeding of cotton increase yields; the wife must be equally involved and provided with an incentive to do the work required. Decision-making in African agriculture can be highly fragmented; and since the husband and wife are not necessarily one economic unit (polygyny brings further fragmentation of economic interests), innovations which require that a husband make the decisions for his wife or wives as well as for himself are likely to fail.

Extension programs are not worth the name if there are no profitable innovations to extend or if the innovations that exist are of doubtful or marginal profitability. Nevertheless, it is clear that as an agricultural system expands its technical base, the need for sound extension advice, supported with profitable inputs, tends to be a critical limiting factor. The challenge for each specific program is to determine when this point has been reached. Like research and other institutional infrastructures that serve agriculture, extension emphasis in tropical Africa has tended to be uneven. Given the dispersal of African farmers over a wide area and the poor road system, the extension program that can be expected to have a development impact is one that has a strong mobile staff which is well instructed in a limited number of crop activities and has profitable innovations to extend.

## The Evolution of Extension Services

Considerable differences in approach, scope, and organization mark extension work in Africa today. These differences notwithstanding, it is possible to distinguish three phases in the development of extension programs in tropical Africa: (1) the phase of limited assistance; (2) the phase of essentially regulatory activities; and (3) the phase of development programs, characterized by active educational and service functions. The first two phases coincide with the colonial period, while the third is more characteristic of the preindependence and the postcolonial period.[1]

### Phase 1—Limited Assistance to Farmers

The major concern of colonial governments prior to 1945 was administration, not social and economic development. Early extension activities grew out of colonial administration and reflected the dominant government priority—the maintenance of political peace. S. M. Makings (1967:63) asserts that a basic principle of colonial governments is *stern* paternalism. Subject people who had no say in the government were required to carry out certain activities—economic and otherwise—that were seen by the administrator as designed to improve their welfare either directly or indirectly.

Early extension assistance to African farmers was limited in scope and in approach. Extension methods were adapted to group instruction in land preparation, seeding, planting, and manuring techniques; soil conservation measures; and in most marginal savanna areas, the enforcement of target "food crop" acreages of the so-called famine-reserve foods.

### Phase 2—Regulatory Activities

In this period, the central axis of colonial government extension programs continued to be crop and livestock production, within a strong

---

[1] A major source of information about extension services in British Africa before independence is provided by G. B. Masefield (1972).

framework of soil and water conservation. The extension approach was changed from the stern paternalism of the earlier period to a benevolent paternalism in which African farmers were not merely told to do what the extension officers considered to be good for them, but were told why it was good for them (Makings 1967:63).

The expansion of an export agriculture in a competitive world market demanded that the quality of the products be made acceptable to consumers. Major emphasis on quality strengthened the regulatory character of the colonial extension service. Regulation of the quality of export crops and control of animal diseases may have been necessary; other regulations, like the requirement of self-sufficiency in food staples in each district and of tie-ridging even when land was plentiful, probably were not and their enforcement was almost hopeless.

Mass immunization of livestock and the control of cattle movements are largely regulatory services; so also are phytosanitary regulations that govern the introduction of new crops and their protection from pests and diseases. In animal industry, a large body of extension activities is rightly regulatory. The control of epizootic disease was a necessary precondition to the establishment of improved and high-yielding stock (but it probably aggravated the distress in the western Sudan when drought became severe). The "command" approach failed on a number of extension activities, whether they were technically impeccable regulations like those cited above or less defensible goals such as commodity "target drives," soil-erosion control measures, and the selling of farm produce to an inefficient government monopoly. Compulsory measures for agricultural improvement were openly rejected before independence, and many technically sound programs fell into disuse once the pressure from government staff was removed.

## Phase 3—Development Programs

Extension work is fundamentally educational. It is based on the belief that new skills and techniques exist that will increase agricultural productivity if farmers will learn them. The expanding government-sponsored development programs which started before independence have tended to increase the hortatory and service activities of the extension service. Whether in an outgrower's system for tea or tobacco, in group farming, in ranching, or in settlement schemes, the extension service has work of educating to do. The more complex the innovation, the greater the need for the educational input of the extension service. On the other hand, the more profitable the new skills appear to farmers, the more willing they are to acquire them. The success of the farmers' training centers (FTCs) in Kenya owes much to the close relation between viable and profitable farm-level programs and the demand for new skills by the farmer. It illustrates the proposition that new farm skills are only acquired voluntarily when they can be used profitably. It should not be inferred, however, that potential profitability will always insure adoption. The problems faced by the early introducers of cotton in Uganda and Zaïre made this clear (cf. Bartlett 1973).

## The Structure and Performance of Extension Services

African governments maintain a monopoly over regulatory services affecting agriculture through their agricultural ministries, but numerous private and quasi-governmental organizations contribute to extension education and service. They include churches, universities, international aid agencies, foreign private organizations, private commercial firms, and farmer cooperatives.

In the decades since World War II, there has been a substantial expansion of government-supported agricultural extension services. In fact, by comparison with the funding of agricultural research, financial support for extension activities has been relatively generous. It is estimated that in 1974 expenditures for extension were equal to about 2.2 percent of the value of agricultural output in Africa. This was nearly 60 percent higher than the level of expenditure for research. By way of comparison, it is estimated that extension expenditures in 1974 in Asian countries represented less than 1.0 percent of the value of agricultural product whereas their research expenditures were equivalent to nearly 1.9 percent. Extension and research expenditures as a percentage of the value of agricultural product differ in similar fashion in the developed countries of Western Europe, North America, and Oceania, but the contrast is even greater. In Europe, for example, outlays for research in 1974 were equal to 2.2 percent of the value of agricultural product but the corresponding figure for extension expenditures was less than 0.6 percent (Boyce and Evenson 1975:8). The relatively high level of expenditure for extension in Africa is, of course, influenced by the fact that the rural population is highly dispersed and by the extensive systems of farming that are practiced. But there is also a suggestion that there has been underinvestment in research relative to extension.

### Administrative Structure

The administrative structure of the extension service, particularly in former British Africa, reflects a unity of command. The service comes under the Ministry of Agriculture, where it tends to occupy an important place. In a typical situation, inherited from colonial days, a deputy director or chief agricultural officer is head of the extension division of the Department of Agriculture (under a director). A senior or provincial agricultural officer is responsible for the departmental work at regional levels, and agricultural officers under him are in charge of districts. In most countries one administrative structure spans the national, regional or provincial, and divisional levels. Nigeria is an important exception, where the state government, rather than the federal, provides the equivalent of a national system of extension service that spans state and district levels. The district is the action level of the extension service, with the responsibility of direct contact with the farmer.

Interaction among the four actors in the extension education system—the farmer, the farmer-training institution, the extension agent, and the researcher and his product—takes place at the district level. Table 7.1 provides an overview of the rank and numerical strength of the extension staff in the seven districts where surveys were carried out.

In 1966, there were 123 extension staff members of all categories in Kisii District of western Kenya, which had an estimated 100,000 farm families, or about 800 farm families per worker (extension workers per thousand farmers is shown in Table 5.5). The ratio between senior staff (A.O. and A.A.O.) and intermediate staff (T.A.) was favorable, about 1:5, as is the ratio of 1:2.5 between intermediate staff (T.A.) and junior staff (A.I.).[2] However, a high staff turnover, resulting from transfers and resignation for further training or other jobs, tended to create a situation of instability, especially among the senior staff. This rapid turnover was also characteristic of Teso District of eastern Uganda. In 1965, there were fifty-one extension staff members of all grades serving in Teso District, one extension worker to about 2,500 farm families. About 20 percent of the total staff were of senior rank (S.A.O., D.A.O., and A.A.O.), excluding staff deployed on special projects and in the farmer-training institutions.[3]

Geita District in Tanzania illustrates how a hastily implemented government program can divert extension resources in ways that adversely affect service to a majority of farmers. At the time of our survey in 1967, the agricultural extension staff in Geita District consisted of two field officers, twenty-one assistant field officers, and thirty-two field assistants.

[2] See Table 7.1 for the meaning of these acronyms. Terminology varies considerably from country to country and over time.

[3] The S.A.O. (Senior Agricultural Officer) in a province or region is sometimes referred to as a Provincial Director of Agriculture (P.D.A). The term District Agriculture Officer (D.A.O.) seems to be universal in the English-speaking countries of tropical Africa.

Table 7.1. District extension staff by rank

| Rank | Kisii (Kenya) | Teso (Uganda) | Geita (Tanzania) | Mazabuka (Zambia) | North Katsina[a] (Nigeria) | Bawku (Ghana) | Akim-Abuakwa (Ghana) |
|---|---|---|---|---|---|---|---|
| *Senior staff* | | | | | | | |
| Agricultural Officers (A.O.) | 1 | 1 | | 3 | | | 1 |
| Assistant Agricultural Officers (A.A.O.) | 6 | 8 | | | | | |
| Field Officers (F.O.) | | | 2 | | | | |
| Technical Officers (T.O.) | | | | 3 | | 1 | |
| *Intermediate staff* | | | | | | | |
| Assistant Field Officers (A.F.O.) | | | 21 | | | | |
| Technical Assistants (T.A.) | 31 | | | 19 | | | |
| Agricultural Assistants (A.A.) | | 19 | | | | 13 | 4 |
| *Junior staff* | | | | | | | |
| Field Assistants (F.A.) | | 23 | 32 | | | | 2 |
| Agricultural Instructors (A.I.) | 85 | | | 25 | | | 3 |
| Commodity Demonstrators (C.D.) | | | | 53 | | | |
| Agricultural *mallams* | | | | | 3 | | |
| Total staff | 123 | 51 | 55 | 103 | 3 | 14 | 10 |
| Farm families per staff | 800 | 2,000 | 2,000 | 200 | 5,000 | 1,300 | 4,000 |

[a]Native agricultural departments advised by Ministry of Agriculture officials are responsible for most of the extension service available to farmers. There were no ministry staff in the survey areas at the time of the study.

This provides one extension worker to about 1,000 to 2,000 farmers who live on dispersed farms, many of them accessible only on foot. In the two seasons prior to our survey, extension work in Geita District had been limited, as a high proportion of field staff had been deployed to provide managerial staff for the mechanized block cotton farms, which were given top administrative priority. However, with the relaxation of the emphasis on the block farm program, because of the poor response of the participants, the extension coverage in the district was being restored.

The experience in Mazabuka District of Zambia was quite different, and it illustrates the conditions under which government priority programs can be staffed without starving a majority of farmers of needed extension services (Anthony and Uchendu 1970:238). In Mazabuka District, the concentration of extension staff was large compared with the other survey districts. In October 1966, the staff consisted of three professional officers, three technical officers, nineteen technical assistants, twenty-five agricultural assistants, and fifty-three demonstrators. (The last category included thirty-two cotton demonstrators and eighteen maize and cattle demonstrators.) The ratio of extension workers, of all grades, to farmers was between 1:150 and 1:200. In both Kenya and Zambia, the use of a large number of specialist demonstrators, who have limited general education but acceptable specialized training in the growing of one crop, tended to make a major impact on production. This suggests one way of dealing with very serious manpower shortage in the short run.

Extension coverage was generally stronger in eastern and central African survey districts than in western Africa, but the level of formal education of the staff of junior grade was lower. In Bawku and Akim-Abuakwa districts of Ghana, the extension staff numbered fourteen and ten respectively in 1967 (.78 and .24 per thousand farmers). In northern Katsina, Native Authority and Ministry of Agriculture extension staff work side by side in the field, much of the actual contact with farmers falling to the Native Authority *mallams* (village-level field assistants).[4] There was one agricultural *mallam* in each of the three survey areas: Dan Yusufa, Zango, and Kadandani villages. Each was responsible for about 5,000 to 6,000 farmers.

## New Programs and Competing Structures

The administrative structure of the extension services in tropical Africa is more complex than the picture that emerges when extension activity is viewed solely as an extension of knowledge to farmers. In many countries

[4]*Mallam* is an honorary title applied to anyone with formal education.

of Africa, specialized development activities have sometimes justified a new system of extension. We have referred to the conditions in northern Nigeria where the thinness of the Ministry of Agriculture staff and political concessions to traditional policies had combined to produce two almost parallel extension systems and two categories of extension staff: the Ministry of Agriculture staff and the Native Authority staff.

The promotion of particular cash crops and commodities like tea and dairy cattle in Kenya, cotton and maize in Zambia, cocoa in Ghana, and tobacco in Nigeria and elsewhere has tended to foster the coexistence of parallel organizations that sometimes make conflicting demands on the farmer. On the other hand, the relatively well-trained and carefully supervised field staff of an organization such as Kenya's Tea Development Authority often obtains impressive results in promoting efficient production of a technically demanding, high-value crop.

## Extension Transplants

A unique feature of the extension system in many francophone West African countries is what may be called "extension transplants." Mali, Upper Volta, Ivory Coast, and Senegal make use of Paris-based organizations, employed on a contractual basis, to promote technical and extension services in specified areas in a number of programs (de Wilde et al. 1967:I, 158–62). Supported by the French government, which sometimes subsidizes their services, the staff of these organizations have tended to be interdisciplinary in character and quite innovative in their extension techniques. The technical and extension services they render are sometimes highly specialized. For instance, the Compagnie Française pour le Développement de Fibres Textiles (CFDT), which operates in Mali, the Ivory Coast, and other French-speaking countries, is a highly integrated single-crop extension system that handles cotton extension work, operates cotton ginneries, and markets the cotton on behalf of the government. The CFDT is an autonomous extension organization that hires and fires its own personnel and is responsible for providing cotton farmers with their necessary inputs—fertilizers, insecticides, sprayers, plows, cultivators—on credit.

The government of Upper Volta has engaged the services of the Société d'Aide Technique et de Cooperation (SATEC) since 1961 to carry out limited extension work focused on village cooperatives. The strategy is to use the village cooperative as a growth point for capital accumulation that would guarantee the purchase of improved farm inputs and implements by cooperating farmers, and as a unit to diffuse technical innovations in the community. Unfortunately, SATEC had no profitable innovation to diffuse and no viable cash crop to offer. Its later programs in Senegal and other countries appear to have been more successful. The Bureau pour le Développement de la Production Agricole (BDPA), working in the Bokoro region of Chad, focuses its extension activity on organizing cooperative groups and on a single crop—peanuts. While BDPA achieved modest success in creating a cooperative marketing structure for peanuts and millet, it has yet to come up with profitable innovations that would justify its educational role.

The Compagnie Internationale de Développement Rural (CIDR), now discontinued, represented a different philosophy of the role of foreign organizations in African agricultural development. The assumption which governed the CIDR's effort was that what limits output of Ivory Coast agriculture is lack of motivation rather than lack of opportunity. The approach was to use a "progressive" villager as *animateur* to stimulate receptivity in his community through his example of becoming the first adopter of technical innovations. Unlike the SATEC or BDPA organizations, the CIDR did not compete with or plan to replace the usual extension service; its basic philosophy was to "prepare the minds" of conservative African farmers to be receptive to future innovations, whether they were designed to promote agriculture or improve rural life in general.

These experiments in the use of foreign organizations in the service of African agricultural development have a number of common features. They use a dense extension network, concentrate on a few areas and on a limited number of farmers, and enjoy the complete confidence of the host government. But they failed to recognize that any institution that wants to serve the long-term interests of a local population must find some way to indigenize itself. An effective development does not thrive on institutional transplants; it requires local institutionalization, which implies the training of local extension workers and supervisors to take over the work.

## The Qualifications of Extension Staff

While an adequate administrative structure and technical support at the national and regional levels are important for the success of extension work, the deciding factor is an establishment at the local level of staff whose training and experience command the confidence of the farmer. The extension service has to compete for available qualified manpower with other employers, and the best graduates are often lost to industry and to the higher administrative grades outside of agriculture. The question is not what the "ideal" qualification for extension staff should be, but the best way to improve and make productive the manpower that can be obtained.

A major problem at the level of agricultural officer is not so much the lack of a university degree as it is the extremely rapid turnover in senior staff. The consequence is a very shaky basis for effective extension work. The intermediate and junior staff are more stable in their postings, but they are less well qualified technically. The range of qualification is enormous. The agricultural *mallams* in northern Nigeria, the *animateurs* in the Ivory Coast, and the assistant agricultural instructors or "farmer extension workers" in Kenya are either poorly trained or self-trained, poorly paid or mere volunteers, and they may render either good or indifferent service to the agricultural community to which they belong.

Agricultural demonstrators, who are usually specialists in one crop or two, are better trained and rewarded than the low-level generalists. In Zambia, demonstrators usually have had eight years of school. Before the season starts, they undergo a specialized course which is a refresher course for existing demonstrators and an introduction for the new intake of staff. Additional courses are provided during the season to highlight various topics associated with the technical operations of any given commodity or crop of specialization. Demonstrators have opportunity for career advancement. After two years of service in the department, they are eligible to take the annual examination for entry into the Monze School of Agriculture, which provides a one-year course designed to produce agricultural assistants. Demonstrators are used extensively in Kenya and Zambia, and have proven quite effective in the introduction of new, high-value cash crops, with supporting inputs and technical services, in areas where economic conditions are already favorable to geographic specialization in agricultural production.

The training of subprofessional or intermediate and junior staff is the responsibility of the Ministry of Agriculture. In eastern and central African countries, assistant agricultural officers constitute the diploma or subdegree level extension personnel. The school certificate is the basic qualification required for professional training in agriculture. Arapai Agricultural College in Uganda provides a three-year course in farm management and farm mechanization, and the Agricultural Training Center at Nyankpala in northern Ghana provides similar training. The Embu Agricultural Training Center, a two-year postsecondary course, represents a typical junior-staff-level training, as does the Ministry of Agriculture Training Institute at Ukiriguru, Tanzania.

## Performance of Extension Field Staff

The actual performance of an agricultural extension service depends on a number of factors in addition to the educational qualifications of its field workers. In fact, a major conclusion of research on the performance of agricultural field staff in Kenya is that problems of organization rather than of individual competence are mainly responsible for the poor performance of many extension workers. Organizational deficiencies which weaken the motivation of field staff, together with inadequate supervision and failure to plan the allocation of work assignments, appear to be more significant than lack of skill or inadequate training (Leonard 1973:4).

Deryke Belshaw and Robert Chambers have studied agricultural administration in Kenya and argue that it is possible to achieve much better performance by field staff by improving the formulation and implementation of agricultural programs (Belshaw and Chambers 1973; Chambers and Belshaw 1973). More specifically, they contend that carefully devised procedures for planning and monitoring the work of field staff make it easier to optimize the allocation of staff time in such a way as to both encourage and require a higher level of performance by field staff. They attach great importance to involving field staff in the development of detailed work programs, arguing that this helps to ensure that targets are realistic and that the staff responsible for implementation have a feeling of commitment to them. Procedures for periodic reviews of progress help to ensure that supervisors will monitor the work of their junior staff as part of an established routine; such procedures also give lower-level staff an opportunity to call attention to shortcomings that are the result of a failure of officials at a higher level to provide necessary support, e.g., in releasing funds. The operation of a management system of this nature appears to have been of considerable value in improving coordination and implementation of the multifaceted programs undertaken in six divisions (subdistricts) in Kenya where experimental Special Rural Development Programs were carried out.

Two additional factors appear to be important for the level of performance of agricultural extension services: the availability of worthwhile innovations suited to the needs of small-scale farmers operating in a variety of agroclimatic zones, and the type of strategy that is adopted for promoting the diffusion of technical information to farmers.

## Role of an Extension Service in the Introduction and Diffusion of Innovations

We have identified the three functions performed by extension services in tropical Africa as regulatory, educational, and service. The educational function is clearly the most important. It is essential to recognize, however, that African farmers still receive most of their technical information and acquire most of their farming skills in informal learning environments. This was especially true of the expansion of cash crops during the first three decades of the twentieth century. During that period, local chiefs and other leaders often played a key role.

Particularly since the end of World War II, agricultural extension services have become a significant channel for the diffusion of knowledge and for providing training in relevant skills. It is not possible, however, for extension staff to reach all farmers, particularly where agricultural production depends on millions of small-scale farmers. Thus the success of extension and related training programs depends basically on the success of such programs in introducing innovations to a relatively small fraction of the farm population and on the degree to which the initial adoption of innovations leads to the diffusion of innovations to other farmers.

The most common approach of extension staff in Africa has been to concentrate on a relatively small number of "progressive farmers" in the expectation that they will set an example and assist other farmers in acquiring new knowledge and skills. Alternatively, the limited extension staff available may be concentrated in a particular area or on a few specialized activities, often a single crop. Along with those common techniques of concentrating resources on a particular group, area, or activity, a number of efforts have been made to reach as many farmers as possible. This type of strategy aimed at reaching the mass of a country's farmers tends to use local leaders and existing social groups or newly created groups. A perennial problem with a mass approach to extension is that when the necessarily limited manpower resources of an extension organization are spread too thin, little or no impact is realized. In recent years there have been some experiments with a promising variation of the group approach. The emphasis is on reaching groups of "average farmers," rather than "progressive farmers," measured by the rate at which they adopt recommended innovations. For reasons that are summarized at the end of this section, such an approach appears to be a more effective means of securing widespread diffusion of innovations and one which also has significant social advantages.

## Institutional Training for Farmers

The demand for institutional training in farm skills is stimulated by a number of factors: government development programs, demand by commercial farm operators and practicing adult farmers, concern with school-leaver unemployment problems, and the introduction of such specialist farm enterprises as poultry management and beekeeping. As John de Wilde

notes, these vocational agricultural schools can succeed only where farming offers profitable opportunities (de Wilde et al. 1967:I, 188–89). The
successful experiences in Kenya, in contrast to the disappointing performance of the graduates of vocational schools in Nigeria and Ghana, illustrate the validity of this proposition. Generally, the training of young men
for direct employment in agriculture—such as tractor and machinery operators for group farms, cooperatives, or state farms—has tended to be more
successful than training them for entry into farming. But the training of
adult farmers, who have a farm to return to and profitable innovations to
exploit, has had impressive success everywhere.

*Training of young farmers.*   The need to provide young people with
productive jobs in agriculture is being recognized by policy makers. While
the view that young men disdain traditional agriculture because of their
formal education still persists, there is strong evidence to suggest that it is
the poverty of traditional agriculture and not something inherent in formal
education that accounts for the flight from agriculture by the young. Young
Africans are willing to train for a career in agriculture and to stay in
agriculture—as long as it is comparatively rewarding.

Agricultural training for young farmers is intended to produce skilled
young men and women who can take up farming in the future either on
their own account, in government settlements, or as agricultural employees
in private or government-related enterprises. Farm schools for young
school-leavers have not had much success in achieving their goals. They
are usually costly programs which can take only a small number of qualified applicants. Except where there is a guaranteed employer for the
graduates, technical training has proven not to be enough to equip a young
man for farming, and the experience of vocational agricultural school does
not ensure a career in agriculture. Where graduate farmers have set up their
own farms, they have had to contend with a host of problems for which the
school gave them little preparation: acquisition of land on a secure basis,
the cost of clearing new land for modern agriculture, and finding an
assured supply of inputs and an assured market for their products.

*Training of adult farmers.*   Adult African farmers are the backbone of
African agriculture. They are in a position to employ productive new
functional skills immediately and to take advantage of profitable innovations and reliable market information. On the other hand, ongoing farm
responsibilities must have the first claim on their time, and the date and
length of the training course must be adjusted to the farm calendar.

The effort and resources that African governments devote to adult farmer
training vary enormously. Our survey data show Kenya to have the greatest

concentration of effort in the area of adult farmer training. The Kenya program of FTCs started in 1955 with three FTCs with a capacity for eighty people. By 1965 there were twenty-nine FTCs with a capacity for 1,500 people, and in 1975 most districts in the country had an FTC.

Although the agricultural and veterinary departments of the Ministry of Agriculture run most of the FTCs, some are operated by church groups like the National Christian Council of Kenya and the Catholic Church, and by government and parastatal organizations like the Ministry of Settlement and the Kenya Tea Development Authority.

The Farmer Training Center, Kisii, illustrates the typical characteristics and changing emphasis of the FTCs in Kenya. This FTC was opened in 1953 as a center for training Kisii farmers in the best way to run their smallholdings, given the new profitable technologies that were becoming available. Each family unit admitted to the center lived on the farm for one year and followed the full agricultural cycle with the management of all crops—cash crops, food crops, and livestock—in the hands of the farmer-trainee under direct supervision by the center's staff. It was envisaged that each holding would provide the farm family with food and would produce about $300 in cash returns, but this assumption proved too optimistic.

The program was modified in the late 1950s, when it became clear that intake was not keeping pace with the rapid expansion in cash crops, especially tea, pyrethrum, and coffee. Few farmers were willing to undertake a long training program at a critical period when the Swynnerton Plan offered them an opportunity to make profitable farm investments. Short, intensive courses, emphasizing specialized farm skills and lasting between one and two weeks, were substituted and had a high payoff. In 1966/67, when our Kisii study was carried out, the FTC had a capacity for 100 trainees and the average rate of utilization was about 75 percent. The factors that made this program successful must be noted. Profitable innovations that justified the time and effort expended in the acquisition of new skills were available. The center attempted to provide the farmer with only the skill and knowledge he needed for actual farm operations. The farmer's wife was involved in the training, and either took the course with her husband or arranged to take it at another time. There was opportunity to update the training and to acquire new skills and knowledge as more profitable innovations became available.

The system of training adult farmers in district farm institutes (DFI) in Uganda is similar to that described for Kenya. There were nine DFIs in Uganda in 1967, all operated by the Ministry of Agriculture and Cooperatives. Courses were of a short duration, lasting one to two weeks. Until

June 1966, the nearby Rural Training Center at Arapai provided short courses for Teso farmers. After it was closed, Teso farmers had to go to either the DFI at Ngetta in Lango District or the one at Tororo. The number of Teso farmers attending the DFI fell from 914 in 1965 to 327 in 1966, showing that many did not consider the training obtained worth the extra distance involved.

Institutional training for adult farmers is better developed in eastern than in western Africa. In Ghana in 1967, the Ministry of Agriculture operated a two-day, Friday-to-Saturday course for farmers. The course introduced farmers to improved farming; among its offerings were poultry keeping, vegetable gardening, nursery management, and tractor driving.

*Specialist training.* A demand for specialist training may be created by the need to use a new technology or by the decision to implement a specialized program. For instance, modern ranching in pastoral areas with limited or poor range demands new skills. The Kenyans who have become large-scale farmers in the former Scheduled Areas require certain skills and management that cannot be obtained from any existing FTC. In West Africa, the acquisition of specialist farm skills revolves around poultry management and the outgrower system for tobacco.

While institutional training for farmers is part of the process of specialization that must accompany economic development, it contributes little to agricultural development in the absence of profitable innovations. Vocational agricultural schools undergo further specialization in the content of their offerings because of demands made on them by the new skills required to make farming more profitable. The Kenya experience shows the dynamics of this system and the critical role profitable innovations play in bringing it about.

## The Elite as a Channel of Communication

Where the advisory service offered to farmers is essentially informational or of a campaign nature, as it is in most mobilization programs, the use of traditional or modern elites in the communication of information can be effective. The successful introduction of cash crops in many parts of tropical Africa was achieved through the use of local African chiefs and village headmen. As long as the new farming technique was simple and the new cash crops could be easily grafted onto the existing farming system, the use of influential local people achieved results. Increased output depended on increased labor commitment and the expansion of acreage, and these could be influenced by the personal intervention of the traditional authorities. Particularly in Uganda and northern Nigeria, the role of the chief or local notable in assuring that the required acreage for cash crops was cultivated was an important one. The use of Native Authority *mallams* as extension workers in northern Katsina represents the institutionalization of the role of the traditional authorities in this communication process.

Political independence has created a new set of leaders who influence the behavior of farmers in ways that can be consistent with receptivity to innovations. In Tanzania, local TANU party leaders, working with leaders of the cell groups (*Kumi Kumi*), are entrusted with the responsibility of mobilizing farmer support for agricultural, cooperative, and general rural development efforts. In Nigeria, particularly in the eastern states, "sons abroad," as Nigerians migrating from their home villages to towns are called, are a major force for change. As sources of pressure, new ideas, and capital, they provide leadership and technical information which improve their home villages (Smock 1969:119). In Teso District of Uganda, the local chiefs have always been an important instrument of extension work. The cooperation of the chiefs was found necessary in the enforcement of certain agricultural innovations, including strip cropping, conservation measures, the minimum acreage requirement for cassava and sweet potato, and other measures backed by local council bylaws. The chiefs still report on crop acreages in their respective areas.

The success of an elite in influencing farmers' adoption behavior depends on the number of farmers who have confidence in them and on their ability to coerce farmers into adopting new procedures. This channel of communication can, of course, be easily abused. The elite have been effective in introducing relatively simple innovations; they tend to be less efficient as innovations become more complex.

Extension programs giving major attention to "progressive" farmers have been most widely used in eastern and central Africa where this class of farmers has often been an important force in rural development. Historically, they have developed through special government programs, they are found where the extension work is and has been active, and they have been privileged groups in their access to institutional resources, training, extension visits, and loans. They tend to be well-to-do, to occupy important positions in the local power structure, and to be receptive to recommended innovations.[5] They differ from the traditional elite in that their status and

---

[5] B. F. Massell and R. W. M. Johnson, who conducted an intensive analysis of farm production data from the Chiweshe Reserve and the Darwin Purchase Area in Rhodesia, concluded that agricultural extension workers tended to rate farmers in accordance with how well they followed instructions rather than on their overall performance (Massell and Johnson 1968:62). Inasmuch as recommended practices have sometimes been based on the technician's concept of "good farming" rather than on an evaluation of costs and returns, progressive farmers are not necessarily superior in terms of their efficient utilization of resources.

influence are due primarily to their responsiveness to new economic opportunities—or to the efforts of governmental officials.

The evolution of the improved farmers in Mazabuka District is associated with the African Improved Farming Scheme that began in the 1946/47 season. The technical requirement for the participating farmer was that he adopt the so-called Kanchomba Farming System, which involved fallow rotation, contour work, soil conservation work, and adoption of reasonable standards of cultivation and weed control. Manure had to be applied on one-quarter of the land, and the farmer had to follow a rotation of maize (manured), maize, legumes, maize. Through a system of price incentives and acreage bonuses, many farmers registered and qualified under the scheme. The number of "improved farmers" rose from 355 in 1952 to 907 in 1959, reaching about 1,410 in 1960, or about 10 percent of the farmers in the survey area (Anthony and Uchendu 1970:228).

The grooming of a special class of "farmer innovators" in Teso District of Uganda started in 1935, essentially as a device to limit Local Authority loans to those most likely to make productive use of them. The scheme was not entirely successful and was abandoned. In the late 1950s, in response to Uganda government policy, a further attempt was made to recruit a class of progressive farmers, defined as those who actively put into practice, in the management of their farms, the advice and instruction of the Agricultural Department. In December 1966 there were 400 progressive farmers in Teso District, most of whom received more than their proportionate share of the available resources, including advisory services, loans, and subsidized farm implements. Because the progressive farmers are a small group, but widely dispersed all over the district, the cost of reaching them tends to be high.

The assumption on which concentration of resources and staff on the small group of receptive farmers is based is that they will act as a positive reference group among their neighbors, i.e., neighbors will copy their superior farm practices. The weakness of this argument is that it ignores the privileged position of the progressive farmer, his access to institutional and technical resources which are not available to his neighbors. The progressive-farmer strategy of extension also runs into political difficulties. Extension resources are subject to political direction and manipulation. (This, of course, is not peculiar to master-farmer schemes, or to Africa.) Insofar as these resources are used to reinforce existing political loyalties, they fail to achieve their primary development objectives.

Given the usual assumption that progressive farmers will be copied by others, the rate of imitation becomes critical, and evidence suggests that it has been slow. These and other considerations have led to a reexamination of the progressive farmer concept, and research and experimentation in Kenya, to which we turn shortly, suggest that new approaches to a group extension strategy may have significant advantages.

## Concentration on an Area

A number of arguments support achieving a high density of extension effort within a particular geographical area. Pilot schemes and experimental programs may call for high-density extension work in the hope that whatever successes are achieved can be subsequently diffused at lower cost in many other parts of the country. Also a unique local situation, including a special physical environment, may call for special attention that requires more than the proportionate amount of resources normally devoted to it, either to correct a situation of distress or to realize the returns from regions of high economic potential.

Complementarities between the development of factor and product markets may make possible much higher returns and much greater development momentum if all the desirable changes can be introduced simultaneously rather than sequentially. Concentration on one area also makes it more likely that the change in the economic environment will be great enough and rapid enough to be easily perceived by the producers. This applies to concentration on specialized activities as well.

The poor performance of African agriculture in terms of needs and political expectations is forcing some countries to explore the impact of resource concentration in agriculture. Starting in 1967, Uganda adopted what was officially called a "saturation approach" as a complementary extension strategy.[6] The aim was to concentrate a large proportion of the available extension staff in a small area and build up an impact through result demonstrations that would be extended to other areas. Thereafter, a gradual diffusion of the successful innovations would be encouraged in other communities. In this model, the concentration strategy assumed the characteristics of a location-specific package. Its rate of diffusion depends on the extent to which the factors that make for its success in the first place can be duplicated in other areas.

As the World Bank has redirected its efforts toward programs aimed at achieving a wider impact, it has often adopted an area approach. A notable example is the Lilongwe Land Development Programme in Malawi.[7] This

---

[6]In Teso District, Miroi Parish in Bukedea County was selected for a pilot testing of this approach.

[7]For a valuable description and analysis of the Lilongwe project and other development projects supported by the World Bank and bilateral agencies, see Uma Lele (1975).

is an integrated rural development program, undertaken with financial and technical assistance from the World Bank and the British government, which has achieved considerable success in increasing farm productivity, particularly by promoting increased use of fertilizers leading to higher maize yields. Although it has had an impact on a much larger number of farmers than was true of most earlier World Bank agricultural projects, the cost per farmer has been too high to permit expansion of the program to a substantial fraction of the country's rural population. This has been partly a result of fairly high overhead costs, but primarily it reflects a strong emphasis on road building, intensive land use planning, health clinics, rural water schemes, and other social services.

The Chilalo Agricultural Development Unit (CADU) in Ethiopia is another noteworthy example of an integrated rural development project concentrated within a limited area. The CADU project received considerable financial and technical assistance from the Swedish government, and once again the cost per farmer reached by the program would make it impossible to replicate that type of program to the extent required to affect a significant fraction of Ethiopia's farm households. An especially interesting feature of the CADU project, however, is the role that it played in preparing the way for a much more massive rural development effort—the Minimum Package Program (MPP), which is being implemented with assistance from the World Bank and FAO as well as the Swedish government. The concentration of resources in the CADU project made it possible to undertake research and experimentation with institutional as well as technical innovations, which greatly facilitated the design and implementation of the MPP. Moreover, by focusing attention on problems of rural development and demonstrating the feasibility of increasing the productivity of the country's small farmers, it also played an important role in "energising the Ethiopian Ministry of Agriculture" to undertake the MPP. An interesting feature of the plans for the latter program is that the number of MPP areas is to be expanded year by year as manpower and other resources are gradually enlarged, so that within a decade a sizable part of the rural population will be reached.[8]

## Concentration on Specialized Activities

Where technology is simple and capital short, as they are in small-scale African agriculture, it might be argued that the opportunity for specialized

[8]For additional details on Ethiopia's Minimum Package Program see Lele (1975:8-9, 203-4). Information is not available concerning the effects of the political changes that

training will be limited. But the opportunities which an expanding market system and modern technology offer for the evolution of specialized agricultural production should not be forgotten in the allocation of extension resources. The Kenya Tea Development Authority's efforts to promote tea growing by smallholders is an important example of specialized programs concentrating on a single crop. Efficient production of tea by smallholders required special arrangements to ensure close coordination between harvesting and processing, and high potential returns per acre justified the creation of the necessary institutional and physical infrastructure.

Experience in the farmer-training institutes has taught that it is a waste of time and effort to overload the content of the curriculum as this tends to leave the farmer-students confused. The need is for short, functional courses that concentrate on one topic that farmers identify as a major farm problem. The use of agricultural demonstrators in those areas where they have acquired specialized skills contributes to the concentration of extension effort. The major contribution to agricultural development made by agricultural demonstrators in Kisii and Mazabuka districts was due in no small part to the fact that they broke the extension tasks into simple functional units that could be learned quickly.

The major criticism of concentration on specific activities is that farming is an integrated enterprise and crop specialists do not often consider farm problems as a whole. But because each farm enterprise is different, it may be that the farmer himself is in a better position to view the totality of his operations than any extension agent.

## Group Approaches to Extension Activity

Because extension staffs are small, various efforts have been made to use social or other groups—village, cooperative, cell, traditional work group, or age-grade—as units of extension activity. The extension worker can also create a classroom away from the vocational or agricultural school by taking the school to the farmers. In Kenya and Uganda, the staff of various FTCs and DFIs have organized "teach-outs" whereby they carry instruction to the farmers. A demonstration plot on a neighbor's farm may provide the object lesson for the teach-out. District farm shows, field days, general meetings to discuss production targets for the districts—all provide opportunities for group extension work.

The experiences of the Village Development Committee in Geita Dis-

---

occurred in Ethiopia in early 1975 on the implementation of the MPP, which was supposed to reach 100 areas, each containing approximately 10,000 farmers, by 1980.

trict, Tanzania, show the strengths and the limitations of one type of approach to extension work. Village development committees were expected to play a very important role under Tanzania's self-reliance program. They had the responsibility to increase production and secure cooperation in self-help projects, but lacked the means. For the 1967/68 farming season, the Geita District Development Committee decided to require every farmer to plant a minimum of two acres of cotton, one acre to receive fertilizer and the other, farmyard manure. The Village Development Committee had the will to implement this decision but not the ability. As a result, farmers more or less ignored the orders of the committee.

An initial pilot extension training project carried out as part of the Special Rural Development Program in Tetu Division of Nyeri District concentrated on farmers somewhat below average in their adoption of innovations. The new approach lessens the tendency for rural communities to be split into "progressive" and "nonprogressive" groups in terms of technologies employed and income. It also appears to be a more efficient means of diffusing innovations to the bulk of the farmers, and to progressive farmers as well as laggards.

Initially, the participants in the Tetu project were selected on the basis of a survey carried out with the assistance of research workers from the Institute for Development Studies at the University of Nairobi. Later, the extension field staff selected participants who had not adopted certain innovations (farmers who had never planted hybrid maize or kept exotic cattle). The participants were taken to the local FTC in groups of about fifty each for a three-day course aimed at persuading them to plant hybrid maize and teaching them how to do so. Steps were taken to ensure that shops in the area stocked supplies of hybrid seed and fertilizer. Participants could get unsecured loans to cover the cost of inputs and the fee for the training course.[9] Eligibility for a future loan depended on repayment, the rate of which has been about 90 percent. Virtually all (97 percent) of the 798 farmers who received training under the program in 1972/73 adopted hybrid maize. Moreover, according to a subsequent random survey covering about sixty farmers, the total number of farmers induced to plant hybrid maize was about four times as large as the number of participants in the FTC training program; and these friends and neighbors of the participants had to finance the new inputs themselves.

This experience has led a number of research workers and extension

[9]An interest rate of 10 percent was charged on the loans, which averaged less than $20 per participant.

specialists in Kenya to reject the notion that concentrating on "progressive farmers" is the most efficient extension strategy. A paper by S. Schönherr and E. S. Mbugua argues that diffusion is slowed down when the extension effort is concentrated on progressive farmers because of the social gap between them and other farmers (Schönherr and Mbugua 1974:8):

> Agricultural innovations are usually rather complex, so that in most cases occasional observations are not enough to enable a farmer to adopt innovations himself. The most progressive have little time or interest for explaining to others in detail the problems and tasks of new cultivation methods. The others do not readily call upon the most progressive farmers because of the social distance between them. Lastly, the average farmers... believe that the most progressive ones are in a position to adopt innovations because they understand all about these things, they have the necessary contacts and are financially secure... so that the agricultural innovations adopted by the progressive farmers are irrelevant for them.

Although the Tetu experiment was viewed as a notable success, it was recognized that the reliance on training courses at an FTC is fairly costly and limits a program's ability to reach many farmers quickly. This conclusion led to a decision to carry out a further pilot extension project in selected areas of Kisii and south Nyanza districts based on training groups of farmers in their own sublocation. Clusters of all farmers in a particular neighborhood were selected, on the assumption that a group of neighboring farmers will be reasonably representative of the average farmers in that area. This procedure proved to be particularly successful when the groups were selected by the villagers themselves during a village gathering, rather than by administrators or extension staff. Concentrating on a cluster of farmers eased transportation problems and facilitated supervision by the instructor. The group met for instruction in the field of one farmer where demonstrations could be carried out. An essential requirement of this approach is careful training of the instructors. Schönherr and Mbugua emphasize that each team of instructors should be prepared specifically for each extension project in a one- or two-week preparation course given in an FTC.

## Service Roles of Extension Work

Early-stage agricultural development may impose responsibilities on the extension service that are not necessarily extension functions. Where the service is practically the only major development infrastructure, it is often forced to assume service roles for which it has neither inclination, training, nor capacity. It is not unusual in tropical Africa for the extension service to assume responsibility for marketing, cooperatives, and even for rural development projects. This is "overextension."

The service role that the extension service assumes reflects the level of underdevelopment existing in the economy, although it may also sometimes reflect a hostility to private enterprise, as, for example, when the extension service is given a monopoly of the supplying of fertilizers. The development process implies specialization of functions, and the more extension is able to restrict itself to its traditional roles, the better the service it can provide. One important way of strengthening the extension effort is to see that the supply of farm inputs and the marketing of farm products are handled by other competent institutions or left to private merchants.

Cooperatives, as an instrument of development, continue to receive attention in many African countries. Their general performance has been disappointing and this has impaired the extension programs by weakening farmers' incentives to innovate. While the extension staff should help farmers to deal with cooperatives in those activities that affect farm enterprises, it is ill-advised for the extension staff to take over responsibility for the cooperatives as this will compromise their relationship with farmers. Assumption by the extension staff of responsibility for collecting debts also impairs their effectiveness.

Extension systems are weakened when they perform roles other than those that contribute directly to agricultural production. Political or ideological propagandizing, distribution of farm inputs, and regulation, while perhaps necessary at an early stage of agricultural progress, tend to divert attention from the central task of extension work during the critical period of agricultural transition. Among the functions that the extension service should phase out to other institutions as agricultural productivity rises are marketing, lending, plant protection, and veterinary services.

## Conclusion

Interest in the valuable role that extension education can play should not completely divert attention from the other ways by which farmers learn how to farm better.[10] We have already stressed the importance of farmers

[10]The nature of extension's role in development and the experience reviewed in this chapter seem to support the following propositions:

1. Extension service is important in the development process but never constitutes a sufficient condition.

learning from one another, a process which has a long history in tropical africa. The spread of manioc across tropical Africa long before Europeans, let alone extension services, had penetrated the continent is a case in point. But maize, peanuts, haricot beans, taro, Asian yams and rice, bananas, and pineapple—to name a few food crops that had their origin outside Africa—spread throughout the continent in a similar way.

Nor should concern about extension education blind us to the important role that private purveyors of farm inputs and buyers of farm products can play in agricultural education. The greatest achievement of tropical African agriculture—the development of the great export-crop industries—is due almost entirely to the probing stimulus of European merchants and the vigorous response of African traders. Private suppliers of farm inputs other than kerosene and gasoline have not been permitted to play as active a role in tropical Africa, primarily because they were mistrusted by the colonial rulers. Where they have, as with the Kenya Seed Company, they have rendered an invaluable service.

Extension education contributes to agricultural development only when there are profitable innovations to extend. In many parts of tropical Africa, the lack of economically rewarding innovations is more limiting than the methods or channels of their diffusion.

---

2. Extension systems that operate without significant new agricultural methods to back their effort have limited impact.

3. The support of professionally trained agricultural scientists who generate or adapt new technologies is therefore crucial.

4. Coordination of research with extension is central if extension work is to respond to farm needs.

5. Extension methods should be adapted to the educational level of the practicing farmers.

6. The more limited the scope of extension effort, the better service it provides.

7. Well-trained supervisors and agricultural administrators and carefully devised administrative procedures for planning and monitoring the work of field staff are central to effective extension work.

8. Rapid turnover of senior extension personnel weakens the whole extension effort.

9. The use of many specialized and well-trained commodity demonstrators with limited formal education is to be encouraged in areas where farming activities are becoming increasingly specialized.

# 8. The Agricultural Research Base and the Flow of Technological Innovations

Publicly financed research institutions to provide a flow of profitable innovations to agriculture are now commonplace in economically advanced countries, but recognition of their crucial role in economic development has been slow in coming. The success of the new high-yielding varieties of maize, wheat, and especially rice has led to a new attitude toward agricultural research in a number of countries and international agencies. This change is apparent in tropical Africa; the creation of an Association for the Advancement of Agricultural Sciences in Africa (AAASA) and the International Institute of Tropical Agriculture (IITA) in Nigeria are manifestations of it. But, generally speaking, there has not yet been a sufficient strengthening of either national or regional (multicountry) research to support programs of agricultural development.

Although the scope and level of support were restricted, agricultural research activities were initiated in the countries of tropical Africa long before these recent events. The resources were extremely limited, but considering the handicaps, a good many significant results were obtained, especially with the major export crops. The paucity of resources devoted to agricultural research was in line with prevailing ideas that colonies should be self-supporting, and the provision of governmental services generally was conditioned by the financial stringency imposed by a limited tax base. The only substantial government outlays were for the building of ports, railways, and roads; the logic underlying those priorities is evident.

The development of some export-crop industries, such as cocoa in Ghana and western Nigeria, was based mainly on informal diffusion of seeds and knowledge and the empirical experience of African farmers in learning how to establish, tend, harvest, and process an exotic crop.[1] The

[1]The crucial role of Ghanaian farmers in adapting and devising the institutional arrangements and technologies that permitted rapid expansion of cocoa in Ghana have been well documented (Hill 1956, 1963). R. H. Green and S. H. Hymer (1966:302) have even argued that the Ghanaian cocoa farmers "succeeded in spite of, not because of, the Colonial Depart-

early introduction of commercial production of cotton, however, owed a good deal to organized efforts to introduce suitable varieties, and it will be seen that cotton was one of the first crops to benefit from organized research. The recent success in expanding smallholder tea production in eastern and central Africa has depended on technical research, as well as on institutional arrangements, for coordinating production and processing.

Although we are concerned here mainly with examining the systematic research carried out by agricultural experiment stations and other institutions, this emphasis should not be permitted to obscure the significant innovative activity of individual farmers and firms. The role of the farmers who pioneered cocoa production in Ghana and Nigeria has already been mentioned. Importing firms and traders probably played the principal role in the spread of simple, improved farm tools. From the beginning of the twentieth century, overseas imports of iron seem to have accelerated the spread of iron hoes and fostered the use of hoes of improved design.

Hand gristmills for maize spread fairly rapidly because of the substantial saving in time as compared to hand pounding. In the Bamenda Highlands, a fairly isolated region in the Cameroons, hand mills were not introduced until 1954, but within four years over 100 maize-mill societies had been formed to purchase mills at the prevailing price of $45 (Miracle 1966: 244-45). In a good many areas, it seems that there has been a direct transition from hand pounding to motorized mills. A recent example involving an initiative by local manufacturers is the development of peanut-shellers in northern Nigeria. These hand-operated shellers have provided a cheap alternative to hand shelling, which they have largely replaced, and because of the saving of labor time and cost they have facilitated the growth of output and increased the marketing of premium-quality peanuts since the late 1950s. Agricultural progress in tropical Africa will be influenced considerably by the strengthening of a creative capacity for such informal innovative activity on the part of individual farmers and of manufacturing firms catering to a rising demand for purchased inputs.

## Development of Agricultural Research Institutions

The availability of technological innovations adapted to the needs of African farmers has been influenced strongly by the historical development of research institutions. Those institutions have a relatively short history,

---

ment's efforts." It seems reasonably clear that organized research did not begin to make a significant contribution to cocoa production in West Africa until the 1930s.

and their work has been handicapped by a lack of continuity of personnel and programs as well as by shortages of funds and of trained scientists and technicians.

The establishment of botanical gardens in the eighteenth century was the earliest attempt to assist agricultural development in the tropics. These gardens introduced and tested new crops and varieties and were often forerunners of colonial agricultural departments and their research services.

European scientists, particularly botanists, had searched for useful and otherwise interesting plants all over the world since the 1600s and had brought them together in botanical gardens. G. B. Masefield (1972:19) calls this the "stream of British experience in tropical agriculture [that was] . . . much the most important one for the future development of [African] government services." Prior to 1700, the only noteworthy British plant collection of this sort was that made by Sir Hans Sloane in Jamaica in 1687/88, but after the development of the Royal Botanic Gardens at Kew, England, which began in about 1759, the idea spread to British territories throughout the world. It reached Africa in 1890. Curators were appointed to Aburi (1890), Lagos (1893), Gambia (1894), the Niger Coast Protectorate (1895), Entebbe (1898), and Sierra Leone (1899) (Masefield 1972:20). In this same period the Belgian government established botanical gardens at Eala and Kisantu in the Congo. In 1902 the German government established an agricultural experiment station at Amani in Tanganyika, which was probably the most advanced agricultural research center in tropical Africa on the eve of World War I.

The experience of the nineteenth century in other parts of the world showed the potential value of crop introductions from country to country. Masefield says that by 1900 this had become "almost an article of faith rather than of policy, and was . . . continued far into the twentieth century." Introductions included cocoa and rubber to Uganda in 1901 and 1903 (the plants came from Kew!), tobacco and vegetables to Somalia in 1905, and Egyptian cotton to Kenya in the same year (Masefield 1972:63–64). Missionaries, trading companies, plantation owners, government administrators, and African migrants all participated in the introduction of plants.

Among the first agricultural departments in tropical Africa were those founded in Uganda in 1900, southern Rhodesia in 1903, and northern Nigeria in 1912 (United Kingdom 1961:7). The first agricultural experimental work commenced in this period and was largely concentrated on export crops. The Royal Niger Company set up experimental stations for investigations on coffee and cocoa in the 1880s. The Aburi Agricultural Station was established in the Gold Coast (Ghana) in 1890. The French established a research station at Bambey in Senegal in 1913 that was focused on research on peanuts and was one of the early examples of the specialized commodity research stations that have been especially important in Africa.[2] Moor Plantation was founded in Nigeria in 1899 as a model farm for the propagation of rubber and to investigate general agricultural improvement. In 1905 the plantation became a center for cotton investigations. In subsequent years it has emerged as one of the more important research centers in tropical Africa, with federal and state experiment stations located in adjacent sites.

The first experimental farm in the Sudan was set up at Shendi in 1902 to explore the possibility of growing irrigated cotton in the northern Sudan and was followed by a second, south of Khartoum, in 1903 and two other stations to test cotton as a rain-grown crop in the southern Sudan (Idris 1969:2). In Kenya, the Scott Agricultural Laboratory was started in 1907; much of its early work was concentrated on coffee. A cotton experimental farm was set up at Kadunguru in Uganda in 1911, and in the same period experimental work was started at Kampala on coffee, cocoa, tea, and rubber (Jameson et al. 1970:3). In central Africa, a general experimental station was founded at Salisbury in 1909.

The 1920s and 1930s saw further expansion of agricultural research, the establishment of some of the major research stations, and greater research specialization. In the Sudan, where a general agricultural research service had been built up in the early part of the century, the Gezira Research Station was started in 1919. The station soon acquired an international reputation for the high standard of its research work, and the Gezira Cotton Scheme profited from the successful application of its research findings.

In East Africa, Serere and Bukalasa experiment stations were established in 1920, and the Amani Station, which had been established in 1902, was reopened in 1927 as a research center for Kenya, Uganda, and Tanzania. This regional station was moved shortly after World War II to Muguga in Kenya, where it became the East African Agriculture and Forestry Research Organization (EAAFRO). A coffee research station was established at Lyamungu in Tanzania in 1933, and a coffee research team was assembled in Kenya a few years later. An agricultural and veterinary laboratory

---

[2]Strictly speaking it was originally a "Station d'Etudes" and did not become a "Station Expérimentale" until 1921 (Gov't. Gen. de l'A.O.F. 1956:5).

was founded in Zambia in 1929. In West Africa, oil palm research was started at Calabar in 1927 and cocoa research at the Moor Plantation in 1930.

A major development of this period was the establishment of the Institut National pour l'Etude Agronomique du Congo (INEAC), a network of research stations in the Congo, with its headquarters in Brussels and a central research station at Yangambi. INEAC experienced tremendous growth between the time of its establishment in 1933 and 1960, when the Congo became the independent state now known as Zaïre. At that time the INEAC station at Yangambi was almost certainly the most important agricultural research center in tropical Africa. Although priority was given to export crops, INEAC was noteworthy in the attention that was given to food crops and to the attempts to devise modifications of traditional systems of shifting cultivation (see Chapter 4).[3]

Much of the early research work in tropical Africa consisted of fact-finding surveys: the classification and mapping of soils, collection and identification of plants, and surveys of pests and diseases. A survey of soil and vegetation types, water supplies, agricultural systems, and development possibilities in northern Rhodesia by G. G. Trapnell and J. N. Clothier (1937) was a noteworthy ecological survey. In Uganda, a first survey of agricultural systems and social activities was carried out by the Department of Agriculture in 1935–38.

Research work was undertaken with limited staff and, in the early 1930s, against the background of a severe depression and consequent staff reductions in the colonial territories. Genetic studies were commenced on some of the economic crops, physiological work was started on cotton and some perennial crops, and studies of the ecology of insect pests were initiated. Fertilizer work was commenced in the 1920s but was largely confined to trials on experiment stations. The emphasis in all areas was on research on the main or possible cash (export) crops. Where a major research program was carried out on a food crop, for example, wheat breeding in Kenya and maize breeding in southern Africa, it reflected the importance of these crops to commercial farmers. Cassava breeding and selection programs between the two world wars were associated with the colonial governments' programs of famine prevention. Agronomic and agricultural

[3]For fifty years two Belgian publications, *Bulletin agricole du Congo Belge* (Belg., Min. des Colonies) and the *Bulletin d'information de l'INEAC* (Belg., INEAC), were particularly important references on agricultural research in the tropics.

economics research received relatively little attention, although there has been increased emphasis on economic investigations since about 1960.[4]

The most notable research results to date have been achieved by specialized research organizations that have focused on a limited number of export crops. One of the more interesting and important programs in the British colonies was initiated in 1921 with the establishment of the Empire Cotton Growing Corporation, known from 1966 until its demise in 1975 as the Cotton Research Corporation. This organization was able to recruit agricultural scientists of high caliber by offering career opportunities that were not dependent on the research programs within a single colony. Cotton research teams were assigned by the corporation to experiment stations in Kenya, Malawi, Nigeria, Sudan, Swaziland, Tanzania, Uganda, and Zambia. The exchange of experience and of planting material among cotton scientists working in different countries was also significant, and was a precursor of similar research networks established by the Belgian government in the Congo and by French scientific organizations in the other francophone territories. In this respect, the work of the corporation also foreshadowed some of the recently established international research centers such as the Centro Internacional de Mejoramiento de Maíz y Trigo (CIMMYT) and the International Rice Research Institute (IRRI) that have made such a major contribution to raising wheat, maize, and rice yields in tropical and subtropical regions during the past decade.

Probably the most spectacular results were achieved by the specialized stations that developed high-yielding varieties of oil palms. The first significant step in that direction was the establishment in 1926 of an oil palm plantations directorate (the Régie de Plantations) in the former Belgian Congo, which was later absorbed by INEAC. The work of the plant breeders at Yangambi in developing high-yielding crosses of oil palms led to significant yield increases in the Congo and stimulated oil palm research in Nigeria and in French West Africa. An oil palm research station set up near Benin City in Nigeria in 1939 was reorganized as the West African Institute for Oil Palm Research (WAIFOR) in 1951. Cooperation between INEAC and a specialized institute set up by the French government for research on tropical oil crops has been especially significant. In 1947 the French Institut de Recherches pour les Huiles et Oléagineux (IRHO) took the initiative in launching a so-called International Experiment involving an ex-

[4]Of the forty-five farm surveys examined by John Cleave, all but five date from 1960 or later (Cleave 1974:11–12). Only W. H. Beckett's (1944) well-known study of Akokoaso in Ghana was carried out before 1950.

change of plant material and technical knowledge between the INEAC station at Yangambi, three IRHO stations in West Africa, and the plantation of the River Financing Company of Johore Labis in Malaysia (Gascon 1968:413). The high-yielding crosses between Asian and African strains that emerged from this program have been a major factor underlying the recent rapid expansion of oil palm production in the Ivory Coast and, on a much larger scale, in Malaysia.

The establishment of WAIFOR and IRHO respectively illustrate important developments affecting agricultural research in the former British and French colonies during the period since World War II. A major recommendation of the influential Hailey Report, published on the eve of World War II, was that the British government should provide substantial funds for the support of research in Africa (Hailey 1938). In addition to WAIFOR, a West African Cocoa Research Institute was established at Tafo in Ghana and a West African Rice Research Institute was created at Rokupr in Sierra Leone. Regional units were also established for maize and for work on storage problems.

In the former French territories, the major emphasis was on specialized research organizations created by the French government, with headquarters in Paris and field programs established at a number of experiment stations in Africa. As early as 1921 an Institut d'Agronomie Coloniale was established at Nogent-sur-Marne, and that organization along with the Musée National d'Histoire Naturelle was assigned responsibility for guiding overseas research programs and for providing specialized training for research workers. Throughout most of the period since World War II, an Office de la Recherche Scientifique et Technique Outre-Mer (ORSTOM) has had overall supervisory responsibility for research in former French territories, but agricultural research has been carried out primarily by specialized institutions dealing with a particular crop or group of crops. In addition to IRHO, with its responsibilities for oil palm, and also groundnuts and coconut palms, reference should be made to the specialized organization with responsibility for cotton research, the Insitut de Recherches du Coton et des Textiles Exotiques (IRCT), and to the Institut de Recherches Agronomiques Tropicales et Cultures Vívrières (IRAT), which has organized research on food crops. Additional specialized organizations have had responsibilities for coffee and cocoa, for rubber, for citrus and other fruits, and there has also been an organization charged with fostering the development and use of improved farm equipment, the Centre d'Etudes et d'Experimentation du Machinisme Agricole Tropicale (CEEMAT).

## Availability of Agricultural Innovations

There have often been significant interactions between technical progress that has raised the yield potential of export crops and the subsequent spread of production among African farmers. As noted earlier, the recent expansion of oil palm production in the Ivory Coast was a direct consequence of the availability of the new, high-yielding varieties and husbandry techniques developed by IRHO.

Cotton offers a number of noteworthy examples of the interaction between technical progress and expanded production. The initial spread of cotton as an export crop was dependent on the introduction of American Upland types. The spread of cotton in Nigeria was based on the Allen variety of Upland cotton, and seed sent to Uganda provided the basis for developing Albar cotton, a variety which incorporates resistance to blackarm disease. Albar in turn has provided the basis for further hybridization and selection in Uganda and in other countries in eastern and central Africa. In the 1940s and 1950s, a small team of scientists working at Ukiriguru Research Station in western Tanzania was able to develop varieties resistant to an insect pest, jassid, and to a disease, bacterial blight, which were the two major biological factors limiting cotton yields. The higher level and increased reliability of cotton yields which resulted from this work, and from the concurrent introduction of seed dressing, undoubtedly contributed to the rapid increase in cotton acreage and output during the 1950s and 1960s. Still more recently, advances in cotton growing techniques have made it possible to grow cotton successfully in Rhodesia, Zambia, and Malawi, where this had previously been impossible because of severe pest problems. The initial technical breakthroughs were the work of a Cotton Pest Research Team working at Gatooma in Rhodesia, and were based on recently developed insecticides and spraying techniques. Similar techniques have permitted successful expansion of cotton production in the Ivory Coast and on a smaller scale in other francophone countries of West Africa.

Among other export crops, spraying to control capsid, the principal insect pest that attacks cocoa trees, has probably had the most significant impact on production. On the other hand, research has not yet provided a means of controlling the swollen-shoot disease of cocoa, except by the costly process of cutting down diseased trees and their contacts, a practice that encounters considerable opposition from farmers even when they receive substantial subsidies for cutting out and replanting. Most of the other important export crops have also benefited to some degree from varietal improvement, improvements in husbandry methods, and better processing techniques such as the so-called wet method of processing Arabica coffee.

Innovations in domestic food crop production have been much more limited, and the important examples can be summarized briefly. Broadly speaking, the new high-yielding varieties of maize, wheat, and rice have had very limited effect in tropical Africa. Maize and rice production have benefited the most. There has been widespread success in obtaining substantial yield increases for maize at experiment stations but relatively few instances of widespread adoption by farmers. The most notable success has been achieved at Kitale in the Kenya Highlands, where recurrent selection methods have been applied to diverse genetic material, including promising varieties from Central and South America where ecological conditions are similar. Both synthetic varieties and varietal hybrids have been developed with excellent yield potential. Considerable attention has also been given to determining optimal time of planting, plant population, and other agronomic practices. A local company has organized an effective program of seed multiplication and distribution, and the agricultural extension service has set up numerous demonstration plots on farmers' fields. The result has been widespread adoption by smallholders as well as large commercial farmers.[5] Promising research has also been carried out to develop short-maturity, drought-tolerant varieties for areas of lower altitude and rainfall, but adoption of these varieties by farmers has been limited. A major program of maize research in Rhodesia has also led to sizable yield increases in that country as well as in Malawi and Zambia, which now have their own breeding programs and, in Zambia, seed production as well.

During the 1950s, plant-breeding programs for rice at Yangambi in the Congo and at Rokupr in Sierra Leone had a considerable impact on rice yields in those two countries, and the more outstanding varieties were introduced in a number of other African countries. In recent years, IR-8 and other semidwarf varieties developed at IRRI have given very high yields under conditions of controlled irrigation, and ad hoc introductions of other varieties from the Philippines and Taiwan have also given good results, although disease problems have caused substantial losses in some

[5]According to a sample survey carried out among smallholders in eleven districts of western Kenya, about two-thirds of the farmers planted hybrid seed in 1973, about half applied chemical fertilizer, and not quite a quarter used manure on maize. Those figures are a weighted average of two high-altitude zones, with adoption rates over 90 percent, and a lower altitude zone with only about 15 percent adoption. This is attributed to the fact that the hybrids available are based on material selected in high-altitude regions and do not have a very strong yield advantage at lower elevations (Gerhart 1975).

years. Rice is mainly grown as an upland crop in tropical Africa, and the wide variation in ecological conditions complicates the problems of evolving varieties and agronomic practices adapted to the various environmental situations. Because of the accumulated knowledge of the genetics of the rice plant and the diverse and promising plant material available, however, the research base for generating worthwhile innovations for rice compares favorably with that for maize. There has been a rapid expansion of rice acreage in various parts of tropical Africa for several decades, and the availability of more productive varieties will surely accelerate that process.

Wheat is of limited importance except for the Kenya Highlands and the high plateau areas in Ethiopia.[6] Wheat research in Kenya has a long history and considerable success has been realized, especially in developing a series of rust-resistant varieties to cope with that recurrent problem. The most interesting recent development, however, has been related to the role of high-yielding, fertilizer-responsive varieties of wheat in some innovative agricultural programs in Ethiopia. The CADU in Ethiopia has rapidly expanded use of improved varieties of wheat and teff and of fertilizer by farmers participating in the project, with an approximate doubling of wheat yields (Nekby 1971; CADU 1971).

Considerable attention has also been given, especially during the past decade, to breeding high-yielding varieties of sorghum, another grain crop with high-yield potential. But results to date have been very limited. In Nigeria, the introduced varieties have a shorter growth period than local varieties, and this has posed difficulties because of the problems of harvesting and storing the crop in the middle of the rainy season. A number of the introduced varieties have also encountered problems of consumer acceptability because they are considered less palatable than traditional varieties, a factor which has also limited the success of a number of maize-breeding programs. In addition, crop losses from bird damage are especially severe for sorghum.[7] Efforts have been made to minimize this problem by reducing the enormous bird populations, especially the quelea or weaverbird, and by breeding for characteristics such as a pendant head (gooseneck) in the hope of making the grain so difficult to take that the birds will feed on other food, such as grass seed. But success may be limited until the evolu-

[6]Wheat does not tolerate humid heat; in the drier savanna regions, rainfall is insufficient and the possibilities for irrigation are limited and costly.

[7]This seems to have been a major factor in the replacement of sorghum with maize in a number of areas where sorghum is basically a more suitable crop because of its considerably greater capacity to produce a satisfactory harvest in spite of periods of drought during the growing season.

tion of farming systems in sorghum-growing areas has reached a stage in which continuous cropping covers extended areas so that there is less shrub and bush growth and fewer trees, thus leading to some decline in bird populations, particularly in relation to the area planted to palatable grain crops (Doggett 1970c:Chap. 12).

Of the other major staple food crops, the efforts to breed mosaic-resistant varieties of cassava has perhaps been of greatest importance. However, this research seems to have been hampered by a lack of continuity in research as well as by the general shortage of funds and manpower. There is undoubtedly a large potential for substantial yield increases for this crop, even without the application of fertilizer or other new inputs.

The low levels of fertilizer use in tropical Africa are to some extent a consequence of the very limited progress in developing high-yielding, fertilizer-responsive crop varieties. There is also lack of knowledge about fertilizer response of major crops grown on various types of soils, with the result that the standard recommendations for fertilizer application frequently are not economical.[8] It is significant that the few localities that have witnessed a rapid increase in fertilizer use in recent years are areas where new, fertilizer-responsive varieties have been introduced. The 15,000 farmers participating in Ethiopia's CADU project, for example, used 5,000 tons of fertilizer in 1971, after only four years of operation; and it was estimated that 9,000 tons would be used in 1972 compared to total national consumption of 7,000 tons in 1969.

The maize farmers in Mazabuka District are another case in point. The increase in fertilizer use on maize in Kisii and most other areas where the new varieties have spread in Kenya has been considerably less; because of the fertility of the young volcanic soils, nutritional deficiencies do not yet significantly limit the yield of maize. Fairly heavy application of fertilizer on a few commercial crops, such as tea, tobacco, and oil palm, accounts for much of the fertilizer that is currently being used, and this again is related to a relatively good research base.

Apart from plant-breeding programs that have increased resistance to disease and pests, plant protection measures have been limited to a few crops and localities. The work of H. H. Storey and R. Leach in Nyasaland

---

[8]It has been determined, however, that application of relatively small amounts of phosphate fertilizer to groundnuts is generally profitable because of the limited availability of this nutrient in the savanna areas where this is a major crop; P. R. Goldsworthy and R. Heathcote have identified four different response zones for northern Nigeria (Anthony and Johnston 1968:33).

during the 1930s represented an early and notable success in plant pathology. Their discovery that a disorder of the tea crop known as "tea yellows" could be cheaply and easily remedied by applying sulfur to the soil made a notable contribution to the later success of that export industry (Masefield 1972:85). Successful spraying with insecticides of cocoa and cotton was made possible by the research of plant pathologists, entomologists, and scientists in related disciplines. For cotton, however, earlier investigations that demonstrated the value of cut-and-burn techniques for reducing pest hazards have been applied much more generally than spraying. Research that has provided a basis for formulating cotton-seed dressings to control seedling diseases, caused by several seed-borne and soil-inhibiting fungi and bacteria, has been especially important because it is relatively easy and inexpensive to treat the seed that farmers obtain each year from a ginnery. Seed treatment has also proved of value for peanuts, and the utility of insecticides for protecting stored peanuts and other stored produce has been established.

Numerous investigations have been carried out comparing sowing dates, spacing, and certain other agronomic practices, and these have frequently provided the basis for recommendations to farmers. An optimal sowing date, most frequently early, is a particularly common recommendation because of the wealth of experimental evidence that yields are reduced if planting is delayed much beyond the onset of the rainy season; but until the early 1960s there was very little farm management research to assess the opportunity cost to farmers of effecting such a change in time of planting. The effect of this recommendation on farm practice has been limited because of the demand for labor occurring at that time, with food crops and export crops competing for the farmer's time. In Nigeria, cotton research workers are trying to adapt their work to relatively late planting of cotton; in Teso District, Uganda, use of ox cultivation has enabled farmers to reduce the time required for planting. Earlier and more thorough weeding is another frequent recommendation by extension staff, but again the effect on farm practice seems limited and for similar reasons. In some areas there has been a considerable increase in plant populations and use of planting techniques that ensure more even spacing. As noted earlier, those practices have tended to accompany the spread of hybrid maize in Kenya and also in Mazabuka; and the use of row markers in Senegal has facilitated closer and uniform spacing of peanuts to minimize rosette disease and the use of ox-drawn cultivators.

Soil and moisture conservation techniques have also figured prominently in extension programs. In many instances those recommendations were advanced as attributes of "good farming" with little regard to their economic value to the farmer. Uganda appears to be one of the few areas where substantial and sustained progress was made in the dissemination of such practices. More recently, however, when techniques to conserve soil and moisture have been promoted as adjuncts to profitable innovations, like the farm layouts and cash crops promoted by Kenya's Swynnerton Plan, the response of farmers has been much more positive.

Mechanical innovations affecting agricultural production directly have generally been of limited importance. The most widespread progress has been made in replacing the hoe with ox cultivation. Mazabuka appears to be exceptional in the extent to which the introduction of the plow has been associated with a range of ox-drawn implements, although there has been extensive adoption of cultivators and planters in Senegal (Monnier, 1975).

Tractor cultivation has been introduced mainly through the establishment of government-operated tractor hire services, but there has been a limited spread of tractors among individual farmers who often do contract work. Because the purchase, operation, and maintenance of tractors is so costly relative to the restricted purchasing power of African farmers, this innovation has affected only a relatively small number of farmers. The emphasis on tractor schemes has often had adverse effects on larger numbers of farmers because the concentration of government funds and manpower on such schemes has sometimes led to the abandonment of programs intended to promote wider use of ox cultivation.

There has been considerable mechanization of auxiliary operations, most notably in the spread of small power-driven mills for grinding maize and other products. But in contrast with the rapid pace of change in Asia, there has been remarkably little use of power-driven pumps for lifting groundwater or water from rivers. This is presumably related in part to the much more limited availability of underground water or perennial watercourses, as well as to the relative abundance of agricultural land so that pressure to intensify farming has in the past not been very strong. However, limited knowledge of water resources and of pumping technologies has certainly been a contributing factor.

Beef production is one means by which natural resources, in this example grassland, can be used to expand farm incomes. A great deal of existing technical information about cattle raising in the tropics can be readily applied or adapted to local situations, and research needs reduced accordingly. Animal health and herd management practices are largely directly transferable and there is a great amount of information on the nutritional requirements of cattle and on management practices for forage and fodder crops. Even where no formal investigations of cattle rearing have been carried out, local knowledge has frequently been acquired through the management of government herds. Increased beef production can be achieved in many areas by the improvement and utilization of local grazing and fodder resources, without the need to resort to the expensive establishment of large areas of planted pastures. The carrying capacity of natural grazing can often be increased by measures such as bush thinning, the use of fertilizer, and the sowing of suitable forage legumes. The introduction of good management practices, for example rotational grazing and the provision of well-distributed watering points for stock, will help maintain grazings in their most productive condition. Natural and improved forage can be supplemented, especially in the dry season, by feeding by-products of the processing industries, such as bran and cottonseed meal, to provide the food requirements for good growth.

## Factors Responsible for Deficiencies in the Research Base

Several of the factors that account for the shortcomings of agricultural research programs in tropical Africa have already been noted. Limited financial support for research during the years prior to independence was an inevitable consequence of the prevailing views that colonies should be self-supporting and that the role of government in economic matters should be distinctly limited. Furthermore, the disruptions that resulted from the world wars of 1914–18 and 1939–45 and the Depression of the 1930s led to serious curtailment of research activities that were already operating with very limited staff and meager budgets.

Several specific problems have limited progress in generating the flow of innovations that is crucial to agricultural progress. Lack of continuity in research programs and failure to make research relevant to actual conditions at the farm level stand out as the two major factors that have compounded the difficulties stemming from limited financial support and lack of trained manpower. Owing to the extreme scarcity of agricultural scientists among the nationals of the countries of tropical Africa, research programs have been heavily dependent on foreign scientists. Even with a considerable expansion during the past ten to fifteen years in the number of African research workers with the advanced professional training required for agricultural research, the supply is small relative to the number required to staff reasonably adequate research programs. Furthermore, technical assistance programs which enable African countries to employ expatriate scientists with broad experience and demonstrated capacity for productive research do much more than simply overcome a quantitative deficiency. Academic training of high quality is a necessary but not a sufficient condition for successful performance as a research worker, and effective leadership in the planning and execution of programs of research is a rare talent that is best nurtured through a "learning-by-doing" involvement in successful programs. Opportunities for that type of experience in tropical Africa have unfortunately been all too limited.

Although the value and need of expatriate scientists is clear, it has entailed important disadvantages. It is almost inevitable that a foreign scientist will lack a deep understanding of existing farming systems and the constraints that condition the ability and willingness of African farmers to accept innovations. In fact, the educational background of many African agricultural scientists is such that not infrequently they also are lacking in understanding of actual conditions at the farm level.

Closely related to the foregoing problem is the lack of continuity in research personnel and programs that has so frequently had adverse effects on the results achieved. It needs to be recognized that an individual with a specialized competence can often make a notable contribution in a short period of time, provided the existing staff are able to convey their experience and understanding of problems to the new staff member and in turn profit from his special knowledge and skills. But all too often the core staff to provide that continuity and fruitful interaction has been lacking. These problems are of course exacerbated when recruitment procedures, the incentives offered, and the conditions of work fail to attract scientists of high caliber.

There is an exceptionally great need for rational determination of research priorities, based on economic as well as technical considerations, under the conditions that prevail in different countries in tropical Africa. Well-conceived research programs can help to minimize the problems of lack of continuity and limited relevance of research findings to farm-level conditions. Furthermore, the allocation of resources in support of agricultural research cannot be justified unless research programs demonstrate their ability to promote increased productivity and output at the farm level, and thereby elicit the political support required for continued and enlarged funding of research.

# 9. Priorities in

# Agricultural Research

The limited financial resources available for research in tropical Africa and shortages of qualified scientific personnel make it important to be able to define lines of investigation that will have the greatest impact in fostering increases in agricultural production in relation to the needs of each country. These needs will be influenced by natural resources and the potential for agricultural production as well as by existing land use, national development policies, and the availability of technical knowledge.

Past experience of successful research programs provides indications of the types of research that are likely to yield large returns under the conditions prevailing in tropical Africa.[1] Key factors in the general acceptance of innovations by farmers are the size of the return obtained and the ease with which an innovation can be understood and adopted. Specific priorities can only be determined within the context of a particular situation. However, the problems commonly encountered in promoting agricultural development within the region make it possible to suggest some general guidelines.

## Financing Research

The need for agricultural research is not directly related to the size of a country or to its production potential. It may be possible to economize in the early stages of research by making use of varietal material and management techniques developed in similar environments. Nevertheless, it will still be necessary to investigate the suitability of materials and tech-

[1]A Committee on African Agricultural Research Capabilities composed of seventeen members from Africa, Europe, and North America has examined research priorities and assessed the capabilities of meeting the research needs of the countries of tropical Africa (NAS 1974). This committee drew upon a large number of papers that were presented and discussed at the 1968 Abidjan Conference on Agricultural Research Priorities for Economic Development in Tropical Africa (NAS 1968) and a number of subsequent seminars and conferences. These included a 1971 FAO conference on the establishment of cooperative research programs (FAO, 1971b) and a series of fifteen seminars in 1970 and 1971 sponsored jointly by the Ford Foundation, IRAT, and IITA.

niques for the areas to which they are to be introduced. For example, if insecticide spraying is necessary for the control of crop pests, information is likely to be available on the chemicals that can be most profitably used. But the pattern of infestation of the crop by pests and their relative importance differ from place to place and need to be evaluated for each situation. The results of such investigations will determine the choice of insecticides and the method and timing of spray applications. Subsequently, the adoption of spraying on a large scale may change the relative importance of crop pests, as a result of the destruction of beneficial parasites and predators, so that the pest situation must be kept continually under review.

Even where knowledge acquired elsewhere can be applied, continued adaptive research will be necessary to solve the problems that arise from changing environmental and economic conditions.

## The Cost of Research

The facilities and equipment required for agricultural research make the cost of maintaining an effective agricultural research worker very much greater than that of maintaining an extension officer. A country with a small national income may find it difficult to support even a modest research effort. Data collected by St. G. C. Cooper (1970) indicate that thirteen countries in East and West Africa for which information was available in the mid-1960s allocated from 0.05 to about 4 percent of their gross domestic product, derived from agriculture, to agricultural research. It is estimated that in 1974 total agricultural research expenditures in African countries represented 1.4 percent of the value of agricultural product. For the countries of Western Europe and for North America and Oceania, the corresponding figures were 2.2 percent and 2.7 percent of their much higher value of agricultural product. It is also noteworthy that in Asian countries, research expenditures as a percentage of value of agricultural product rose from 0.6 to 1.9 between 1959 and 1974, whereas the increase in the African region was only from 0.8 to 1.4 percent (Boyce and Evenson 1975:8). The more rapid increase in support for research in Asia has been especially pronounced since 1965, no doubt reflecting the influence of Asia's "Green Revolution" on policy makers' perception of the economic value of agricultural research.

The shortage of national research workers at the time of independence of former colonies presented an even greater problem than shortage of finance. A general scarcity of professional manpower made the problem acute, nor could the agricultural research services provide the same opportunities for well-paid employment and advancement as other government services and business. Individual countries have made substantial progress in training their own research personnel, both in their own universities and overseas, but research services still depend to a large extent on the supply of experienced foreigners. In nine countries in East, central, and West Africa for which data are available for the mid-1960s, only four (Guinea, Liberia, Nigeria, and Sierra Leone) employed a greater number of national than of expatriate agricultural research workers (Cooper 1970). A few countries had to rely entirely upon expatriate scientists.

## Overseas Aid

Without assistance from foreign countries and, in some instances, their own private sector, it would not be possible to maintain present research efforts in many African countries. Resources allocated to research are supplemented by international agencies, foreign governments and organizations, international foundations, notably Ford and Rockefeller, and by industry. Private industry makes important contributions in some countries. The tea and pyrethrum industries contribute to research in Kenya; the sugar industry in Mauritius supports sugar research through a cess on exported sugar; chocolate manufacturers finance a research unit working in West Africa; and in Swaziland cotton breeding and seed maintenance are largely financed by a crop levy self-imposed by cotton farmers. Throughout tropical Africa, scientific staff are provided under various technical assistance schemes and their salaries wholly or partly paid by the donors.

An overseas aid program that has a strong extension bias will contribute little until there is something to extend that is clearly rewarding for the farmer to adopt. This is commonly lacking and can only stem from research. For this reason, it is desirable that bilateral and multilateral technical assistance programs give increased attention to augmenting the financial and personnel resources available for agricultural research in tropical Africa. The possibility offered by research aid projects for training local graduate counterparts by experienced expatriate scientists also needs greater recognition.

## Local Support

Local investment in research will depend on the extent to which local political support is generated. This support will, in turn, depend to a considerable extent on the success of research organizations in developing innovations that have a substantial impact on production at the farm level and on their ability to attract and retain competent national scientists.

The Sudan provides a good example of what can be achieved. Agricultural research commenced at the beginning of the present century. It has been of a consistently high standard and has played a valuable role in development of crop production, particularly in the production of extra-long-staple cotton, the mainstay of the Sudan's economy. The control exercised over farmers on large-scale irrigation schemes helped the acceptance of recommended farm practices and pest-control measures. The issue of new cotton varieties with a high degree of disease resistance ensured a readily marketable product and provided insurance against heavy crop losses in years favorable to the spread of bacterial blight. Government's recognition of the contribution made by agricultural research is demonstrated by substantial investment in research and in the training of research personnel. Because of the prestige of the research services, it is not difficult to attract recruits. In 1954, prior to independence, there were 32 agricultural research scientists, all expatriates. By 1968, the fiftieth anniversary of the founding of the Gezira Research Station, the Agricultural Research Corporation of the Sudan employed 128 graduate staff members, of whom 81 were fully qualified specialists. All except 8 were Sudanese (Idris 1970).

## Organization of a Research Program

The technical requirements of agricultural research and the special circumstances of the developing countries suggest guidelines for the establishment of national research programs. Because resources available for research are so limited, major emphasis should probably be given to local adaptive research to investigate the applicability of innovations under a variety of local conditions. The developing countries cannot afford the luxury of research that is largely academic.

In the organization of a research program, decisions have to be made on the allocation of scarce resources among disciplines and among the most promising fields of research. Investigations that are designed to improve the interacting components of a farming system demand teamwork which it is not always easy to achieve.

### Teamwork

One of the basic characteristics of agricultural research is its dependence on a wide range of scientific disciplines. Crop improvement cannot be effected within the confines of a single discipline. In developing varieties resistant to pests or diseases, the plant breeder must work in close cooperation with entomologists or pathologists. A change in variety may necessitate a reappraisal of cultivation practices and the cooperation of agron-omists. Or a change in methods of cultivation may make a change in plant structure desirable and require the cooperation of the breeder. Al-though teamwork is essential for successful agricultural research, it is often hard to accomplish and the need for an interdisciplinary approach makes it difficult to draw up plans for research and to determine the priorities to be given to projects.[2]

The success of agricultural research programs must be judged by the extent to which recommended innovations are adopted by the farming community and by their effect on farm production. Attention has to be paid to the difficulties experienced by small-scale farmers in adopting new farming practices under constraints imposed by limited resources. There is as much need for a flow of information from farmer to scientist as from scientist to farmer.

In the developed countries, informed farming communities are not slow to make their needs known. The unorganized small-scale farmers of de-veloping countries are less likely to bring their problems to the notice of agricultural department staff. Difficulties traditionally experienced in farm-ing are often regarded as part of an accepted way of life, and not identified by farmers as problems needing solution.

The agricultural problems that face the developing nations can only be solved by integrated research programs involving both natural and social scientists, working to common ends. The need for team effort presents problems in the establishment and maintenance of satisfactory balance between the personnel required by the various disciplines and for the spe-cific needs of commodity research.

## Single-Crop Institutes

Single-crop institutes staffed by interdisciplinary teams working at sta-tions specializing in specific crops have played an important part in tropical crop development. There is a strong case for their establishment for crops like the perennials that can be studied largely in isolation or for crops like tobacco that require special treatment. International institutes of this sort can carry out specialized work that supports and is often beyond the means of national research programs. Where research on a valuable crop is neces-sary to safeguard the interests of a portion of the farm community, a specialized research unit can be financed most fairly by a crop levy. How-

[2]I. Arnon (1968) provides a particularly valuable discussion of these and related issues with special emphasis on the problems confronting less developed countries.

ever, the provision of separate facilities for several crops in a national research service can lead to the wasteful use of limited funds and duplication of effort. It can also lead to scientists losing sight of the need to regard their work as part of an integrated effort to improve total farm output. The coordination of work on specific crops can best be achieved by the creation of teams, with agreed programs, working within the framework of the various disciplines making up research departments.

## Long-Term and Short-Term Research

Related problems arise in establishing a balance between research directed at solving immediate problems and research designed to meet long-term needs of crop production.

The decisions that have to be made in improving methods for the control of cotton pests in tropical Africa are typical of those necessary in drawing up research programs. The cotton plant is particularly vulnerable to insect pests, and continued effort has to be put into improving methods of pest control. At present, emphasis is placed on the use of insecticides, linked with a closed season during which the crop must not be grown. This necessitates routine screening of the new insecticides that continually become available and studies of the methods and the timing of their application to the crop so that appropriate advice is given to farmers. Such work is time-consuming, and where only one or two entomologists are available to work on cotton pest problems, it occupies a large proportion of their time. Nevertheless, biological investigations are also necessary in order to provide the basic knowledge needed both to improve current methods of control and to develop alternatives. These investigations include work on the biology of pests, their flight patterns, the time of infestation and the extent of damage to the cotton crop, seasonal food preferences, factors determining host resistance, and natural control of the pests by parasites and diseases.

It is difficult to justify basic biological work until an insecticide program has been established. But once this is done, careful assessments should be made of what further work can be attempted and how a balance can best be struck between the demands of the shorter-term insecticide work and the longer-term, but complementary, biological investigations. Similar decisions have to be made in allocating research priorities in other disciplines and crops.

The wealth of the countries of tropical Africa is so heavily dependent on farming that it is especially important for agricultural research to be directed to the highly practical problem of improving the productivity of farmers. By reason of their facilities for research, the universities can make major contributions to national agricultural research programs by encouraging their staff to undertake the more fundamental investigations.

## International Cooperation

The cost of research programs can often be significantly reduced by exploiting the opportunities that exist for the international exchange of ideas, information, expertise, and plant material. Exchange is frequently helped by international organizations. CIMMYT and IRRI have provided breeding material, technical aid, and training facilities to support food production programs in Latin America, Asia, and Africa. Similarly, the former British Cotton Research Corporation provided information, advice, and practical help on cotton growing to a number of developing countries. This aid was provided mainly through a team of scientists who cooperated with overseas governments in research programs designed to improve the yield and quality of cotton.

The former colonial powers, which still have considerable expertise in tropical agriculture, have established home-based research institutes with the object of providing highly professional technical assistance to the developing countries. In the United Kingdom, the Ministry of Overseas Development has a number of major scientific units engaged in research and surveys in the renewable natural resources field: the Centre for Overseas Pest Research (COPR), the Tropical Products Institute (TPI), the Centre for Tropical Veterinary Medicine (CTVM), the Directorate of Overseas Surveys (DOS), Land Resources Division (LRD), and the Overseas Department of the National Institute of Agricultural Engineering (NIAE). These units carry out investigations for and provide technical advice to overseas governments as well as providing training for overseas research workers and technicians. Similarly, seven institutes have been set up in France under ORSTOM to undertake research in tropical agriculture. They include IRAT and separate institutes for research into oil crops, rubber, coffee and cocoa, and fruits; IRCT provides services similar to those of the Cotton Research Corporation.

A need for easy exchange of information was recognized as early as 1928 when eight bureaus in selected branches of agricultural science were established in Britain to provide clearing houses for information in agriculture and forestry throughout the Commonwealth. There are now fifteen Commonwealth bureaus and institutes in the United Kingdom, which collect and disseminate information throughout the world.

The growing amount of research that is carried out in the developing countries makes it increasingly difficult for scientists to keep in touch with developments in similar fields of work elsewhere. Wasteful repetition of research effort will result unless those concerned with research programs familiarize themselves with the progress of research work in other countries and the results of earlier work carried out by their own organizations. Liaison between research workers is often inadequate, even between those working in the same field, and it is not unusual for there to be little contact between separate research institutes within the same country. Liaison is essential to avoid fragmentation of effort and to ensure the efficient use of scarce resources. The publication and exchange of registers of research programs and of up-to-date progress reports on recent work are two of the ways by which information can be made more readily available than at present.

## Regional Cooperation

Because of variation in rainfall and altitude, many countries cover a wide range of environments. A country as small as Burundi (27,800 square kilometers) has four distinct physiographic zones. Each zone has its distinct agricultural problems, and commercial farming ranges from the production of rice and cotton under irrigation at lower altitudes to cattle grazing and barley, peas, and potato farming at higher altitudes. In Kenya, farming takes place at altitudes from sea level to about 7,000 feet. Several West African countries are traversed by a series of ecological zones, from rain forest in the coastal areas to semiarid country in the north. Many countries find it necessary to carry out investigations for a range of ecological conditions, each zone presenting its own special problems. It would be logical for countries with common conditions and crops to pool their resources in the establishment of research institutes with responsibilities for work in common zones or in common crops. Effective cooperation among the states concerned would enable the establishment of large integrated teams, beyond the resources of individual states. The cost should be less than that resulting from uncoordinated efforts.

Stations of the West African Oil Palm, Cocoa, Rice, and Stored Products Research Institutes were given regional responsibilities during colonial days in British West Africa. Since independence, the various African regional research institutes have become national institutes. A notable exception was the East African Agriculture and Forestry Research Organization (EAAFRO), a regional institution serving Kenya, Tanzania, and Uganda.

Financial support from outside East Africa enabled it to cope with a number of serious problems caused by the weakening of the East African Community. Late in 1977, however, those problems resulted in a breakup of the common services of the East African Community, and EAAFRO has now become the Kenya Agricultural Research Institute. International institutes cannot usually be expected to prove workable unless financed and directed by an independent donor. Examples are IITA in Nigeria, the International Center for Tropical Agriculture (CIAT) in Colombia, and the International Crop Research Institute of the Semi-Arid Tropics (ICRISAT) in India, in which the Rockefeller and Ford Foundations are collaborating. In recent years this international network of research centers, including IITA, has also been receiving significant assistance from the World Bank and from the bilateral aid programs of the United States, Canada, West Germany, the United Kingdom, and other countries.

The establishment of regional research programs on an ecological basis is a possible alternative to regional institutes (Devred 1968). The first objective would be the comparatively modest one of promoting cooperation in research and coordinating action on a limited number of common problems. Experience has shown that a lot can be achieved on the basis of cooperation between individual scientists and with a minimum of official links. The AAASA, created in 1971 with a secretariat based in Addis Ababa, offers promise of facilitating such cooperation. The gains from regional cooperation could be great and the fostering of this cooperation would seem to be a useful way in which the Organization of African Unity (OAU) could contribute to the well-being of its member states. The OAU's Scientific, Technical, and Research Commission has already taken responsibility for a major regional research program on cereals and a second program aimed at applying the results of research in a campaign to control rinderpest (NAS 1974:173).

## Economic Considerations

The potential contribution of agricultural research to national output will depend in part on the present production and prospective demand for the commodity that will be affected. Returns are likely to be especially great for research directed at important commodities for which effective demand is increasing rapidly. A relatively modest increase in the yield of a major crop can result in a large increase in total production, although farmers are more likely to adopt a change that increases their own returns significantly.

In order to maximize the economic returns from applied research, it is important to establish sound criteria for determining priorities in selecting research projects. Farm management studies which lead to the identification of the most important technological constraints that are limiting the productivity and output of farmers in various areas can provide valuable guidance in identifying the types of innovations that are likely to be most profitable (Belshaw and Hall, 1972; Vail, 1973). Research priorities should, of course, also be guided by judgments concerning the prospects for success. The emphasis on seed-fertilizer innovations in the next section is related particularly to the strong probability that the selection of high-yielding, fertilizer-responsive varieties will prove to be a cost-effective means of increasing production. A later section stresses the potential contribution of improved equipment because of the evidence which indicates that shortages of labor, especially seasonal labor bottlenecks, are often an important constraint on output both through their adverse effects on crop yields and on the area cultivated. In brief, there are cogent reasons for developing both land-augmenting and labor-augmenting innovations, provided that the practices and new inputs that are recommended are within the reach of small-scale farmers.

Expanded output of a processing industry may also be a consideration in deciding research priorities, especially when increased supplies of an agricultural raw material are required for its operation. Textile mills are a case in point. The last twenty years have seen the establishment and rapid growth of national textile industries throughout tropical Africa. The special requirements of domestic textile mills are likely to influence cotton research programs. For example, until recently the whole of Nigeria's cotton crop was exported and a single variety was grown for this purpose. The need for a quantity of better-quality cotton by the growing Nigerian textile industry, which in 1971 absorbed 217,000 bales, led to the development of a longer-staple type. This new type is now in commercial production in the eastern cotton region of the northern states.

Research has to be oriented to the requirements of local industries; it is desirable for plant breeders to receive early advice of likely changes in requirements of the processing industries, as it takes ten years or more to develop a new variety of an annual crop. With greater emphasis on production of edible oil and livestock feed, the cotton breeder may be expected to produce seed with high oil and low gossypol content. Similarly, the development of a canning industry will create a demand for varieties of fruit and vegetables with color, shape, size, and canning properties not required by local consumers.

## The Seed-Fertilizer Approach

A set of innovations that probably should receive top priority in most if not all African countries has been emphasized by Arnon. These are techniques which, "when applied in combination, give rapid and spectacular results in a very short time. A combination of improved variety, appropriate fertilisation, adjusted plant population, efficient weed control and plant protection can give increases in yield that range from fifty to several hundred percent" (Arnon 1968:92). The advantages of this type of research strategy have been borne out by experience in the countries that have been exploiting the potential of high-yielding, fertilizer-responsive varieties of maize and the new semidwarf varieties of rice and wheat. The maize programs in Kenya and Zambia are important examples in tropical Africa.

Packages of techniques are most likely to be understood and accepted by farmers if they are relatively simple. It was postulated earlier that the acceptance of new innovations is cumulative and that the success of one or two innovations will stimulate acceptance of subsequent practices. It is desirable to keep the number of innovations initially recommended to a minimum. One of the easiest methods of obtaining a substantial increase in production is by the introduction of potentially high-yielding varieties that are responsive to fertilizer application.

Where there is close control over seed distribution to farmers, as with cotton throughout tropical Africa and wheat in Kenya, plant-breeding programs can make a rapid impact on national production. It is easy to introduce a new strain, although it may not be immediately obvious to the grower that it is better than the previous strain issued to him. The concept of good and bad seed and of the response of crops to manure is generally understood by farmers, however, and helps the adoption of new varieties and fertilizer, even when there is no control of seed issues.

Both seed and fertilizer can be adopted on whatever scale fits a farmer's purse or creditworthiness. The farmer can, if he wishes, plant only a small field as a trial. The successful example of a few farmers can initiate acceptance of the new innovations by their neighbors and lead to cumulative adoption by increasing numbers of farmers. Such innovations pass from farmer to farmer without outside assistance, and can be expected to find increasing adoption even where extension services are thinly staffed.

Substantial returns obtained as a result of the adoption of a new variety or the use of chemical fertilizer give farmers the confidence they need to try other recommended practices and farm enterprises if the essential complementary factors are easily available. In dry areas, the full benefit of investment in seed and fertilizer will be achieved only if plant water needs are met; but the cost of providing water will have to be balanced against the costs of investing in other means of increasing production. The alternatives include crop selection and development of farm practices that make effective use of available soil moisture and also investment in infrastructure and other measures to facilitate settlement of areas with more reliable rainfall.

## Plant Breeding

Advances in genetics and plant-breeding techniques and the possibilities of international borrowing of promising plant material increase the prospects for fairly rapid progress in breeding, thus lowering the cost of developing new varieties. Major breeding programs on all the important African cash crops and on many of the food crops, particularly the cereals, provide valuable sources of advanced breeding material. Although non-cereal food crops have received less attention, material can be obtained from several national programs and from IITA and ICRISAT. The tradition of ready exchange of material between breeders is still observed, except where commercial interests are involved.

Field and laboratory tests may show introduced material to be suitable for direct introduction into commercial production. Frequently, however, exotic varieties do not prove suitable for direct issue to farmers although they may have desirable characteristics that merit their inclusion in a national breeding program. For example, American Upland cotton varieties are usually not suitable for growing as commercial types by small-scale farmers in tropical Africa, but they may be used in developing locally adapted varieties that are earlier and have a more compact plant shape, larger boll size, or higher ginning percentage. For this reason, in cotton-breeding programs throughout Africa, use is made of American-bred varieties as well as material obtained from other national breeding programs. Thus, the cotton variety Albacala was derived from crosses made in Uganda between an African Albar and an introduced American variety.

In a similar way, high-yielding exotic varieties are used in sorghum breeding throughout the continent, although they are frequently unsuitable for direct issue to commercial growers. Attempts to develop higher-yielding varieties for northern Nigeria provide an example. Local sorghum varieties are tall and late maturing (taking five to six months from sowing to harvest), and maturation, which is controlled by a photoperiodic response, takes place at the beginning of the dry season. Only about 10 percent of the total dry matter produced is grain, and the average yield from the region is only about 600 pounds of grain per acre (Curtis et al. 1963). Numerous short-statured varieties have been introduced, and many give high yields at experiment stations. Unfortunately, the grain quality of these varieties is unacceptable in the region, they mature during the rains, and many are susceptible to leaf diseases. When they are sown late, so as to mature in the dry season, yields are reduced. However, there is promise of combining the desirable characteristics of local and exotic material, as well as using local short-statured mutants, to produce short, late-maturing varieties with high-yield potential and acceptable grain (Andrews 1974).

Plant-breeding techniques are largely transferable and most countries will find it desirable to rely on outside sources for the development of more effective breeding methods, enabling them to devote scarce professional staff to immediate crop-improvement problems. More efficient breeding techniques have become available in recent years. Wide seasonal and locality differences within the extensive areas for which plant breeders in tropical Africa are usually responsible emphasize the need for good adaptability in seed issues in order to stabilize production.

In recent years it has been possible for maize breeders to benefit from advances made in population genetics and studies of the comparative efficiency of various breeding systems. Thus a maize improvement program initiated in Kitale, Kenya, in 1955, and later strengthened by the addition of a genetics unit, started with the production of synthetic varieties. This was followed by the production of hybrids through double and triple crosses, but emphasis has now changed to the use of recurrent selection to improve composite material with high genetic variability (Ogada and Allan 1968). Faster progress may be achieved by the use of recurrent selection methods of breeding on broad-based composite breeding populations than by the production of hybrids. Yield increases of about 5 percent per year have been obtained with composite material, compared to about 3 percent in the conventional hybrid program (Harrison 1970). There are further practical considerations. Seed of open-pollinated varieties is not as costly as that of hybrid varieties and does not need to be renewed every season. There are substantial reductions in the yields of the second and subsequent generations of hybrid varieties. Moreover, synthetic varieties, usually having greater genetic variability than hybrids, are less likely to be as specific in their environmental requirements. The very significant advances made in maize breeding in Kenya were achieved through the availability of a wide range of introduced material and by making use of recent advances in quantitative genetics to develop a suitable breeding scheme.

Sorghum breeding in Africa is proceeding along lines similar to those of maize, although important differences between the two crops affect the procedure adopted. The sorghum crop is largely self-pollinated, whereas maize is cross-pollinated, and the small size of sorghum flowers make them difficult to emasculate. However, the identification of lines carrying cytoplasmic male sterility makes it practicable to produce hybrids for commercial production, while work is also in progress on the use of recurrent selection in sorghum populations in which crossing is ensured by the use of genetic male sterile material. Population breeding methods developed for outbreeding crops are now being adopted for a largely self-pollinated crop (Doggett 1970a, 1970b).

In cotton breeding, there has been a significant change from the exploitation of long-established material to the greater use in hybridization programs of exotic material from outside the continent.

The progress of the maize, sorghum, and cotton breeding programs in Africa emphasizes the important part played by the international exchange of plant material and information. This not only lowers the cost of research by significantly reducing development costs but also reduces the time taken to achieve objectives.

New varieties will be more effective and will be more readily accepted if plant characteristics are tailored as closely as possible to the requirements of the farming system. The period from sowing to maturation should be suited to the environment and the cropping sequence. The structure of the canopy is important for weed suppression and determines the efficiency both of light interception and of penetration by insecticides. In the long run, it will also become desirable for plant structure to be adapted to any special needs of mechanical cultivation.

Cotton provides an example of how the farming system can affect breeding objectives. Varieties of cotton developed in tropical Africa usually have an indeterminate fruiting habit, a characteristic well adapted to the conditions of small-scale farming. By means of an extended growing season, plants are able to compensate partially for loss of early formed fruiting points caused by insect damage or drought, and establish a worthwhile though later crop if end-of-season growth conditions are favorable. These varieties are less satisfactory when the limitations imposed by unfavorable soil and water conditions are removed. Growth tends to become rank, so that the crop is difficult to manage, and yields are reduced by boll rot and as a result of less effective pest control. A change in farming methods may therefore frequently necessitate a change in crop characteristics. In the case of cotton, improved farming methods create the need for varieties with compact plants and a more determinate fruiting pattern.

Plant physiologists can contribute to the development of better crop varieties. A better understanding is needed of the complex relationship between crop yield, morphological and physiological factors, and the environment. An improved root structure may increase the availability of water and nutrients to the plant and enable crops to survive dry periods. Differences in shape and orientation of leaves affect light interception and photosynthetic activity as well as the ease with which insecticide sprays penetrate the crop. Such changes have, in fact, been part of the change in "plant architecture" which has been a major element in raising the yield potential of the semidwarf varieties of rice and wheat. Crop yields could be greatly improved if the means became available to control the direction in which the products of photosynthesis are translocated within the plant and to lower the ratio of useful to nonuseful plant tissues. These problems require a long period of investigation which will be practicable only at a few large institutes, but their solution will help plant breeders to identify and select the type of plant best suited to a particular combination of conditions. Such work may also lead to the development of more suitable cultivation practices. For example, low maize yields are a problem of the lowland tropics. Experimental yields of up to 6,000 kilograms per hectare are obtained for maize in lowland areas of tropical Africa and compare with as much as 12,000 to 18,000 kilograms per hectare in trials in mid-altitude areas of eastern and central Africa. Recent work at IITA indicates that soil temperature is a major factor limiting maize yield and suggests emphasis on minimum tillage of maize fields under lowland conditions.

## Fertilizers

The use of fertilizers as a substitute for the long fallow period enables land resources to be used more efficiently and reduces the amount of labor required for land clearing and preparation. The effect of fertilizers can be outstanding when applied to the new high-yielding cereal varieties. They respond better to added fertilizer and to substantially higher rates of application than do the local unimproved varieties. Moreover, economic responses to fertilizer are not confined to improved varieties. Much of tropical Africa has highly weathered soils that are poor in plant nutrients, and local varieties of crops can usually be expected to respond to moderate fertilizer applications. A better understanding of the relationship between soil type, the nutrient content of the plant, and fertilizer applications will enable fertilizer recommendations to be made confidently on the basis of soil or leaf analysis, and this may reduce the need for large numbers of simple fertilizer trials to be carried out on farmers' fields, to enable recommendations to be adapted to local needs.

The prices paid by African farmers for fertilizer are often high compared with world prices because of high costs of transportation and distribution. However, the latter costs can be expected to be influenced by the volume of sales and to account for a declining proportion of the farm-gate price as sales increase. This is a further argument for area specialization in agricultural development.

The use of small fertilizer granulating (prilling) units offers the possibility for further savings on fertilizer costs. Such units can provide African countries with no national fertilizer industries of their own with the means to make up the range of formulations required for different crops and soils. In some areas, sulfur and minor elements, for example boron, may be in short supply and can profitably be added to the formulation provided to farmers. Because of leaching, more attention needs to be given to the possible advantages of using fertilizers like sulfur-coated urea and concentrated phosphate, which have a high content of nutrients that are slowly released to the soil, thereby increasing the efficiency of their use by plants. Urea, with more than twice the nitrogen content of ammonium sulfate, can significantly reduce the farm costs of nutrients—with the reservation that the use of urea alone can lead to sulfur deficiency problems. It is desirable to introduce high-analysis types before large numbers of farmers become accustomed to using low-analysis fertilizers. Extension advice to farmers should be kept as simple as possible, and where several nutrients are needed, they are best supplied in a compound fertilizer.

## Reducing Risk

The small farmer often cannot afford to take risks beyond those that are inevitably associated with farming. He is often inhibited from investments in improved production methods by the very real possibility of serious crop loss, or even failure, arising from drought, insect pests, or diseases. Anything that will help him to bear the risk involved in farming will hasten the speed of change. The breeder can help by developing crop varieties that are able to give at least a tolerable return in dry years and are resistant to the more damaging pests and diseases.

### Crop Water Needs

In the tropics the seasonal distribution of rainfall determines the cropping season of rain-grown crops. Over large areas of tropical Africa total annual rainfall is low and the amount received highly variable. Evaporation rates are high and crop failure from drought is a common hazard. Studies of potential for irrigation development will receive high priority in countries where shortage of water limits development and possible sources of irrigation water exist. Costs of water development are high and a full assessment of the results of such work must include an appraisal of the returns to similar investment in rain-grown areas.[3] But other methods of reducing the risk of failure because of shortage of soil moisture are also available.

The amount of water transpired by a crop depends on the amount of water available to it, the weather, the stage of development of the crop, and the period it occupies the ground. In areas with distinct wet and dry seasons, depth of the rooting system is a major factor controlling the availability of water to crops. The root development of short season crops in the tropics has received little attention, and E. W. Russell (1967) identifies it as a subject requiring research. Modification of the crop root structure by breeding and the encouragement of deeper rooting by suitable management practices could improve resistance to the periodic drying of soils in both the savanna areas and the humid tropics.

Total crop water requirements can be reduced by the development of varieties with short growing seasons. This has been done in order to improve maize yields in a mid-altitude area of Kenya where rain falls in two seasons, each with a mean rainfall of about twelve inches, and the period of reliable rainfall is limited to about sixty days. Local varieties of maize, the main food crop of the area, are late maturing, and fail in years with low rainfall. In response to this situatioh, varieties were produced at Katumani Experiment Station that flower while there is usually enough moisture in the soil to bring the crop to maturity, and give a yield of 500 to 700 pounds of grain per acre in seasons when the traditional varieties succumb to drought (Ogada and Allan 1968; Harrison 1970).

Agronomic research can also lead to the more efficient use of limited rainfall. For example, a radical departure from conventional cotton-growing methods has made it possible to reintroduce cotton into areas of the central and eastern provinces of Kenya which previously lacked a cash crop. Separately, neither of the rainy seasons of the area had enough rain for a satisfactory cotton crop. However, good yields are possible when cotton is sown just prior to or at the beginning of the short rains (October–December), the period during which the vegetative framework of the crop is initiated. There follows a dry period in January and February when the crop receives no attention and growth is at a standstill. With the advent of the long rains, from April onward, growth is resumed and the

[3]Market gardening is a special case; where dry periods affect stability of production, supplementary irrigation of high-value vegetable crops is usually worthwhile.

effective crop is harvested from fruits developed in these second rains. Further research should determine what constitutes a desirable plant structure at the beginning of the long rains, how it is best achieved, and the best timing of field operations. Under such conditions, there is also a need for varieties that are able to recover from drought conditions and compensate for the early loss of fruiting points. Thus a less determinate growth habit and smaller bolls are more desirable than would be the case if an adequate and reliable supply of water were available to permit normal crop development.

A great deal of information already exists about farm practices that make the most effective use of soil moisture by crop plants. These include means of reducing surface runoff, water harvesting,[4] conservation of soil moisture, and prevention of erosion. The adoption of suitable cultural practices and varieties enables good use to be made of limited rainfall. The best combination of practices will vary from region to region, and adaptive research is necessary to determine how technology developed for farming in other dry regions can be transferred to specific African conditions.

## Pests and Diseases

H. H. Cramer (1967) has estimated that perhaps a third of the value of all potential world crop produce is lost to pests and diseases during growth, harvest, and storage. Because of the lack of evidence, such a figure can only be a guess. There is no doubt, however, that these losses are substantial.

New pest and disease problems are likely to arise as a result of the adoption of new crops and varieties and more intensive cropping methods. The cultivation of increased areas of genetically uniform material—for example, a popular, high-yielding cereal—can lead to an increase in pest and disease incidence, as can the adoption of near monoculture practices and the greater use of fertilizer. In traditional farming, sowing a mixture of crops and crop varieties, which are themselves heterogeneous, gives protection against devastating loss from pests and diseases. Crop losses vary from season to season and depend upon climatic factors and the extent to which farmers are prepared to carry out control measures.

The adoption of chemical control practices is usually slow. Farmers often find these measures difficult to understand or are deterred by the cost of chemicals and equipment. In order to spray a crop, it may be necessary

---

[4]The collection of surface runoff to supplement soil moisture in an adjacent area lower in elevation.

to carry large quantities of water to the field in areas where water is in short supply. Other factors that cause crop prospects to be poor can make pest-control measures by themselves uneconomic. Under these circumstances, the plant breeder makes a valuable contribution by the development of varieties with good resistance to the major pests and diseases. The use of resistant material provides a cheap and easily understood method of pest control and is the easiest means of stabilizing and increasing crop yield. Resistant varieties require no action from the farmer beyond the initial purchase of the correct seed, and in one-variety areas, where only seed of a chosen variety is made available, the question of farmer acceptance does not even arise.

The development of cotton production in Tanzania shows what can be achieved. A damaging leaf-sucking insect, the jassid, and a disease, bacterial blight, severely reduce the yield potential of susceptible varieties in Tanzania in some seasons. The issue to farmers of seed of varieties which gave improved yields, initially in the presence of jassid alone and later of both jassid and bacterial blight, greatly reduced the risk involved in growing cotton and undoubtedly contributed to the expansion of cotton production (Arnold 1970). Commercial cotton production in Tanzania increased from an average of about 48,000 bales of lint a year in 1946–51 to a level of about 400,000 bales in the 1970s. This increase was achieved both by an increase in the areas sown to cotton by farmers and by an increase in yield from about 130 to 240 kilograms of lint per hectare.

Constant watch has to be kept for any breakdown of crop resistance. This may be the result of the emergence of new races of pathogens. Or a race that had been comparatively rare may become more common. Where resistance is specific to a few races and determined by one or a few major genes, it is not likely to last long. The most efficient resistance is nonspecific and is usually inherited by a polygenic system (Hooker 1972; Howard et al. 1969).

Resistant varieties cannot be developed rapidly and, though best suited to the resources of the small farmer, are only one means of pest and disease control. An integrated system using a number of control methods—resistant varieties, chemicals, and cultural control—is likely to prove most effective. This implies continued biological investigations to determine how losses resulting from the pest-disease complex can be most economically reduced. In the foreseeable future, reliance will have to continue to be put largely on chemical control and will necessitate the routine screening of available materials and testing of new application equipment and techniques. For reasons given above, chemical spraying is not likely to be quickly accepted, but research leading to reduction of the cost and effort incurred in spraying can be expected to lead to increased rates of adoption.

The use of seed dressings is another means by which risk can sometimes be reduced. Potential loss to the crop arising from damage by seed-borne diseases and soil pests can be significantly reduced at low cost to the farmer in money or effort. Where applicable, dressing seed is one of the most effective and economic methods of chemical control. Where the farmer purchases his seed, the operation may be carried out on his behalf, as is commonly done with cottonseed in tropical Africa.

## Improved Equipment

Most cultivation in tropical Africa is still carried out with hand tools. Seasonal shortages of labor—particularly at the time of crop establishment, early weeding, and harvest—limit the areas of crops grown and the quality of crop husbandry. Inadequate weeding is a common cause of low yield. A farmer's ability to adopt new techniques often depends on the possibilities open to him for overcoming labor shortages. The use of mechanization to reduce labor needs at critical periods can be a key factor in development (Vail, 1973). Within each farming area there is a range of farm size and farmer prosperity. This creates corresponding needs for a range of cultivation equipment from hand seeders and push-type weeders to relatively sophisticated ox-drawn multipurpose tool bars, as well as a demand for contract plowing services.

A wide range of machinery is available for small farms, particularly ox-drawn and light spraying equipment (Boshoff 1968; Labrousse 1968; Minto and Westley 1975; Darrow and Pan 1976). But information is not readily accessible to extension services on the types of equipment that are available and on the suitability of equipment for specific tasks and conditions. It is best supplied by government testing centers, working in cooperation with manufacturers, responsible for development work and in contact with international advisory services such as the National Institute of Agricultural Engineering in England and CEEMAT in France.

Research on mechanization problems should be done by interdisciplinary teams responsible for evaluating the need for specialized equipment and the probable consequences of mechanization, as well as for determining the most suitable equipment. Emphasis should be put on machinery that is relatively uncomplicated and, as far as possible, capable of local manufacture—for example, ox-drawn equipment, small tractors, and simple processing machinery, including shellers, manioc graters, threshers, grinding machinery, oil-expressing equipment, and peanut and fiber decorticators.

Tractor mechanization will play an increasingly important role in the agricultural development of these countries as the predominantly agrarian structure of their economies is transformed. The rate at which such equipment is introduced should, however, be geared to a real need. It must be supported by the provision of adequate servicing facilities and the means for farmers to acquire the necessary expertise to enable them to manage their machinery to the best advantage. Sharply increased fuel costs and higher prices for tractor equipment underscore the necessity of a cautious approach to tractor mechanization, and may well help to focus more attention on ox-drawn implements.[5]

Numerous makes of small knapsack spraying machines are available at a wide range of prices, but shortage of clean water prevents their widespread adoption. About 220 to 330 liters of water may be needed to spray a hectare of a field crop with conventional knapsack sprayers. The range of small spraying equipment can be further improved by the development of machines that apply ultra-low volume (ULV) sprays of two to five liters per hectare of insecticide formulations. Because the use of water is eliminated, labor costs are substantially reduced. Battery-powered ULV spraying machines, weighing very little and easily carried by farmers, are now available. Special oil-based formulations recommended for use with these machines are more expensive than the water-based formulations used in conventional spraying, but recent research indicates that the cheaper water-based formulations can be successfully substituted and costs of spraying further reduced. The development of lightweight, inexpensive, ultra-low volume equipment should contribute to the more rapid diffusion of adequate crop-protection measures and make spraying practicable in drier areas.

[5]There may be a place for a small tractor of about ten to fifteen horsepower, suited to the small size of African farms. Small equipment has less demanding technical requirements than equipment conventionally used on medium- and large-scale farms and could be manufactured locally, with only limited reliance on imported components. However, the low traction efficiency that is a consequence of their light weight appears to put these small tractors at a cost disadvantage as compared to conventional tractors of thirty to thirty-five horsepower or larger, used for contract plowing. In an attempt to improve the mechanical efficiency of small traction engines, the National College of Agricultural Engineering of the United Kingdom has revived an approach used in California farming in the early years of this century. The tillage implement is drawn across the field by a self-propelled winch attached to a cable anchored at the edge of the field. In this way, power is transferred directly to the implement without loss through wheel slippage. The college is carrying out testing and development of its "Snail" in Britain and Malawi (Muckle et al. 1973; Crossley and Kilgour 1974).

## More Productive Farming Systems

The characteristics of the developing countries of tropical Africa suggest further criteria relevant to the determination of research priorities. If programs are to be effective, they must take account of the full range of action required to achieve a significant increase in yields and production among large numbers of farmers. Developing countries can ill afford research which results only in an increasing gap between the yields observed on experimental stations and those obtained by farmers. The full implication of the adoption of new farming practices must be assessed so that appropriate action can be taken. Consider, for example, the matter of timeliness of carrying out cultivation operations.

In tropical areas with a single rainy season, an annual crop must be sown at the beginning of the rainy period if it is to yield enough to justify its cultivation. Crops sown early benefit from the mineralization of organic nitrogen and the flush of soil nitrate in the early rains, before it is lost by leaching, and are less likely to suffer water stress later in the season. Soil temperature may be a major factor responsible for the effect of time of sowing on crop yields in some areas. The productivity of an annual cash crop will suffer if, as commonly happens, farmers give priority to the sowing of their food crops. A practical solution to this problem will depend on success in obtaining greater productivity from farm labor and on the willingness of farm families to rely on local markets for increasing amounts of their food supply. However, the design and widespread use of the type of improved farm equipment discussed earlier is almost certainly part of the answer. Timely farm operations which are essential if good commercial yields are to be attained may require major changes in traditional crop cultivation. Introduction of improved equipment for land preparation, sowing of crops in lines in order to make weeding easier, and the use of herbicides are some of the means by which labor requirements can be reduced at critical periods. Practical answers to problems such as these can only be obtained through a flexible approach to research by broadly based research teams working closely with extension staff.

Integrated crop-improvement programs can provide a focus for coordinating the often fragmented activities of research and extension. They must combine research aimed at the improvement of varietal material and methods with the action needed to persuade farmers to adopt new methods. To be fully effective, such programs must include training programs for extension staff and activities needed to ensure easy availability of improved seed, fertilizers, pesticides, and the other inputs needed, and the feedback of information from farmer to research worker. Involvement of research workers in the whole range of crop-production problems appears to be the most effective means of ensuring that agricultural research is practical at the farm level.[6]

The special need for more productive alternatives to the prevailing systems of shifting cultivation has already been discussed in Chapter 4. Under shifting cultivation, soil fertility falls and weed growth increases during the period that the land is cropped. During the subsequent bush fallow, plant nutrients are restored to the higher horizons of the soil profile, via the vegetation, and weed growth is suppressed by competition from other species. Increased population pressure, expansion of the areas used for cash as well as food crops, and the greater areas that can be cultivated by plow farmers lead to shorter fallow periods. It is desirable that at least a few research centers in tropical Africa, representative of the main agricultural zones, carry out investigations aimed at developing more intensive and more productive farming systems.

Studies initiated by INEAC in the 1930s, a first serious attempt to modify existing agricultural systems, were prompted by the failure of efforts to transfer European agricultural practices to the former Belgian Congo. Cropping sequences, the durations of periods of cropping and bush fallow, and farm layouts were established for the main environmental types, with modifications for local conditions. The *paysannat* system was an attempt to rationalize indigenous farming systems and to facilitate the adoption of better methods of land preparation, pest control, and crop storage and processing where land was plentiful, and to prepare the way for the adoption of new methods. Those efforts came to a halt at the time of independence, and problems associated with the *paysannat* system that were mentioned in Chapter 4 have apparently not been resolved. The

---

[6]There is an urgent need to evolve more efficient research methodologies for generating technical innovations suited to the needs of small farmers. The interdisciplinary research carried out by the Rural Economy Research Unit (RERU) in northern Nigeria represents a promising approach to this problem (Norman, 1974). Another promising approach has been developed at the Instituto de Ciencia y Tecnologia Agricolas (ICTA) in Guatemala (Hildebrand 1976). An agrotechnical and socioecnomic reconnaissance survey of existing farming systems and constraints is carried out in order to understand what practices farmers in a particular region are using and why. This information provides the starting point for determining research priorities and objectives. Initially priority is given to identifying simple innovations that are feasible under existing conditions. Trials carried out by research staff on farmers' fields help to verify the suitability of innovations, and subsequent "tests" of selected innovations by representative farmers provide further verification of the profitability and feasibility of innovations to be promoted by coordinated action by the staff of regional research stations and extension workers.

development of improved farming systems is a major objective of IITA, although that aspect of the institute's research program is still at an early stage.

A good many isolated attempts have been made by agronomists elsewhere to rationalize mixed cropping systems. Mixed cropping has undoubted merits. It is a traditional means of obtaining early ground cover of cropped land as well as of spreading farm labor requirements and risk.[7] It is often possible to obtain a greater return from a suitable combination of crops sown as a mixture, than from the components of the mixture sown in pure stands. However, mixed cropping systems are less appropriate for the adoption of new techniques. Herbicide use is difficult, and it may not be possible to spray one crop because of the possibility of unacceptable insecticide residues in the other crop with which it is grown. A great deal of field experimentation is needed to determine the best pattern of mixed cropping. It is justified, however, where there is reason to expect that it can result in a substantial increase in crop production per unit area and where precise spacing and timing of operations is not essential for satisfactory results. Cereal-legume mixtures, such as maize and beans, offer the most promise. Moreover, with alternate-row planting, many of the advantages of mixed cropping can be obtained while still making interrow cultivation possible.

The development of soil-management practices leading to the improvement and conservation of soil fertility must be an integral part of the work on farming systems. The use of fertilizers enables soil fertility to be maintained or improved in the absence of a bush fallow. In soil fertility studies at Namulonge in Uganda, the cycle of plant nutrients in the profile during an arable-grass rotation on red ferralitic soils was measured to obtain the information needed to formulate a suitable fertilizer mixture for the area. Routine use of the mixture over twelve crop seasons (six years) arrested the decline in fertility of land under arable crops (Jones 1968b, 1971).

Under some conditions, use of a planted fallow, possibly containing a legume—for example, *Stylosanthes*—may be of value, especially if it realizes an economic return by providing forage for livestock. But it is likely that continuous cropping will be achieved by the application of sufficient amounts of balanced fertilizer to sustain heavy crop yields. However, much more needs to be known about the effect of cropping sequences on nutrient levels in the profile, the best means of maintaining good soil cover, and the most efficient means of controlling weeds. Weed growth

[7]Belshaw and Hall (1971, pp. 55–56) list nine potential advantages of intercropping. See also Norman (1974, pp. 31–32).

will increase as the fallow period is shortened or eliminated, and special attention will have to be given to studies of species such as *Imperata cylindrica, Cynodon dactylon,* and *Cyperus* spp. that have proved difficult to control by cultural and chemical methods.

Many countries in tropical Africa badly need to develop more productive farming systems for their semiarid regions. Some of these areas, especially in East Africa, are experiencing rapid population growth, resulting from migration from crowded high-potential areas, as well as natural increase. Crop yields are low and risk of crop failure is high. Increased human and animal populations have led to overgrazing, overcultivation, and deforestation by woodcutters and charcoal makers.

Finally, means must be found to further acceptance of new systems by farmers. Acceptance will be most ready if it can be linked to profitable and feasible sets of innovations which constitute components of the improved farming systems yet can be adopted sequentially.

## Adoption of New Methods

The acceptance of new innovations by farmers tends to be cumulative. The successful adoption of one new and rewarding practice strengthens readiness to innovate. Consequently, agricultural research needs to be related to the requirements for sequences of innovations that are adapted to widespread adoption by small-scale farmers with limited cash income. Thus, a new high-yielding crop variety, of acceptable quality, will probably be adopted fairly readily and rapidly. Adoption of fertilizers is more likely if farmers are using a crop variety that responds well to them. But it may be advantageous to introduce a seed-fertilizer combination and in this way to increase the visibility and profitability of the change. The adoption of insecticides is more likely when farmers have a valuable crop to protect. The successful adoption of practices leading to high yields of one crop will prompt acceptance of other new varieties and enterprises. It may, for example, as in the Kisii Highlands of Kenya, subsequently lead to investment in improved livestock, better animal husbandry practices, and acceptance of the need for land enclosure.

At an early stage of economic development, farmers are more likely to adopt innovations that result in easily apparent increases in output, employ divisible inputs, are simple, and have reasonably certain consequences. To the extent that an innovation meets these requirements, it is likely to be self-propagating. Farmers may adopt unprofitable innovations under exhortation or duress, or by mistake; they will not persist in them long if they are free to choose.

Particular attention should be given to the identification and elimination of causes for slow acceptance of new practices. These will often change with each new innovation, but their recognition and solution are essential for the integration of research findings into commercial application.

Attention to the above criteria can help in finding the most rewarding balance among the numerous agricultural research problems of most African states. The success of research organizations in developing innovations that have a substantial impact on production will, of course, determine whether the investment in research yields high returns. Moreover, the ability to generate local political and financial support and to attract and retain competent scientists, both national and expatriate, will depend to a large extent on success in translating research findings into increases in productivity and output at the farm level.

## Summary

If agricultural productivity is to increase, it is not enough that farmers stand ready and able to take advantage of changing economic opportunities, that product and factor markets be efficient, and that agricultural extension services reach a large number of farmers: the ways in which productive inputs are combined, and the inputs themselves, must be altered and the technical frontier pressed outward. This will only be possible if a continuing search for new materials, new plants, new machines, and new methods can be maintained. The research base for sustained development of tropical African agriculture is narrow, but the opportunities are great. It is of critical importance that a high priority be assigned funds needed for continued and widened agricultural research in national development plans.

# 10. Some General Propositions and Afterthoughts

General statements of the requirements for economic progress of small-holder agriculture, whether based on African, Asian, or American experience, are remarkably similar. This similarity occurs regardless of the academic discipline from which the problem is approached. The "secrets of success" of Chapter 3, drawn from an essentially economic analysis of our case studies, are not unlike the statement about sociocultural forces determining innovations that concludes Chapter 6, and both are compatible with Guy Hunter's "inexorable laws" as they were set forth in Chapter 1.[1] They revolve around profitability and feasibility, and they depend on access to markets, to technical knowledge, and to productive inputs.

It is gratifying to find various students of agriculture in different parts of the world arriving at such a consensus. Unfortunately, agreement at so high a level of generality abstracts from the specific considerations that must determine policy decisions. It conceals the tremendous variability in immediate problems from country to country and from place to place. To transmute these wise pronouncements into specific instruments for change requires a great amount of detailed study of local conditions, as the preceding chapters illustrate.

## Hypotheses and Results

In Chapter 1 we set forth a list of questions that we felt needed to be examined if more effective strategies for agricultural development were to be designed and implemented. It is hoped that these questions helped to guide the reader through the chapters that followed and to facilitate his or her appraisal of the usefulness of the findings. We cannot pretend to have answered all the questions that were raised earlier—some have been barely

---

[1]They are also similar to the essentials for development and modernization of agriculture set forth by Arthur Mosher (1966:183)—markets for farm products, new farm technology, local availability of farm supplies and equipment, adequate incentives for farmers, and transportation.

touched on—but it will be worthwhile to review quickly some of the answers that are suggested by our studies.

## Technical Questions

Clearly there is a potential for increasing farm productivity by introducing new methods that require modest investment of scarce resources and are appropriate for small farms. The difficulty lies in identifying these methods and in a lack of the research required to test potentially promising innovations and adapt them to local conditions. Our case studies revealed a frequent shortage of new techniques that were both feasible and productive. Highly productive alternative technologies do in fact exist, but, with a few outstanding exceptions, they have not been adapted to local conditions or made generally available to farmers.

The introduction of new, higher-value crops for export has dominated agricultural change in the past, and most technical change, too, has been related to these crops. Innovations in the production of domestic food crops have been much fewer. Agricultural research suffered earlier from the general view that colonies should be self-supporting; even more debilitating were the effects of two world wars and the Great Depression of the 1930s. Since World War II, research has been handicapped by lack of continuity and frequent lack of relevance to farmers' requirements, both resulting in large part from the extreme shortage of local agricultural scientists. As a consequence of deficiencies in the research base, the methods recommended by extension staff frequently have offered little in the way of increased returns to farmers, and have tended to be discontinued once pressure for their adoption is relaxed.

It was not possible to explore the question of variability of technical efficiency among farmers as thoroughly as we would have desired. Evidence from the case studies and from other farm enterprise studies indicates that variation may be large, suggesting that the scope for increasing productivity by dissemination of methods of the best farmers may be considerable. The matter has been little studied. However, it is necessary to review in detail how the practices of good farmers differ from those of most farmers before the likelihood of their successful propagation can be determined. In the eight study areas, physical resources and farm size showed little relationship to farm productivity.

## Sociocultural Questions

The social and cultural characteristics of African societies seem not to pose serious barriers to economic or technical change. Typically these characteristics have changed as the economic and technical order has changed. There is an interaction between the social, cultural, economic, and technical systems; a change in one stimulates a change in another that may stimulate change in a third. It is often difficult to identify cause and effect. Are the Kisii of East Kitutu progressive farmers because they belong to the Seventh Day Adventist Church, or did they join that church because joining gave them greater freedom to depart from traditional practices? Are inheritance practices changing in some matrilineal societies of West Africa because technical changes have led to increased income and larger inheritable estates? Or is it a change in the economic order that keeps young men at home and fosters a close relationship with their fathers instead of the closeness to a mother's brother that characterized an earlier generation? Or does the change in inheritance rules provide incentive to farmers for heavier investment in their farms? Some lines of causation seem clearer, however, particularly those from changes in the economic and political order to increased availability of farm labor.

It is a peculiarity of African societies that each possesses a common holistic culture in which differences in status are not associated with different levels and kinds of cultural traditions—"great" and "little" traditions in Robert Redfield's terms (Redfield 1956). The differences in world view between townsmen and countrymen are slight. This phenomenon is paralleled by a lack of correspondence of rural with farmer and urban with craftsman. This valuable unity of culture eases the dissemination of new ideas and the functioning of markets. It has been weakened by the growth of new cities that are in the European tradition and by the direct transfer of development models from European and Asian societies.[2] The importance of the extended family, often cited as a shackle on the ambitious, proves in fact to be a source of strength. Kinsmen often supply farm inputs, including credit, and money for education. In many societies the family provides constant reinforcement of the desire to achieve wealth and status. It is apparent that in these societies, as in others, a farmer's interest in increasing output is sustained by the approval of his peers.

Farmers in all of the study areas appear to have been relatively well informed about the practices that were being recommended by the agricultural extension service. There is some evidence that farmers in Bawku and Wanjare are less aware of new methods than farmers elsewhere, but the evidence is not firm. It seems reasonably certain that lack of farmer aware-

[2]Even the term *peasant,* now widely used in East Africa, is considered by many to be inappropriately applied to African farmers. The Belgian agricultural development program in what is now Zaïre was in fact designed to *create* an indigenous peasantry—a *paysannat indigène.*

ness is not nearly so important an obstacle to technical change as the inadequacies of the recommended practices themselves. New crops that offered farmers clear economic advantage have spread through tropical Africa with little or no assistance from extension services. The history of adoption of new techniques and crops in Mazabuka, Kissii, Teso, Katsina, and Akim-Abuakwa confirm that when a practice is seen to be good, farmers learn about it quickly from one another.

Our studies did not examine the consequences of relying on methods other than exhortation and market forces to induce farmers to adopt new methods. That market forces—including improved transportation—are effective is quite clear. Evidence about the long-term effects of compulsion is mixed. In Uganda, for example, compulsion seems to have been important in introducing cotton, one of the country's principal export crops. On the other hand, soil-conserving practices like tie-ridging have generally been abandoned once compulsion ended.

## Economic Questions

Factor markets did not receive special attention except as they appeared to constitute serious barriers to changes. Clearly the extreme fragmentation of holdings that is found in some of the very densely populated parts of tropical Africa—Kigezi District, Uganda, is an example—increases the cost of production and supervision. But the separation of holdings into several separated plots that may be ecologically distinct seems not to be very inefficient, except when animals are kept. In such instances, it is important that the corrals be under direct observation to protect them against predators or theft. It is noteworthy, however, that the three districts showing particularly lively change at the time of our studies—Mazabuka, East Kitutu, and Geita—were areas of recent settlement or resettlement.

Traditional land-tenure practices are frequently cited as a major barrier to change. We found them not to be, except possibly in Bawku, where conservatism is reinforced by a patriarchal family structure.

Income and expenditure records obtained from the case studies and from various household expenditure studies consistently show money income to be above, sometimes, well above, the amounts that are required for taxes and certain "essential" consumer purchases. The implication is pretty clear that many farmers could make modest purchases of new inputs like fertilizers, pesticides, and seeds out of their own funds if they thought it advantageous to do so. They can also borrow from kinsmen. Costly purchases, like dairy cows, trained oxen, and ox-drawn implements, place a greater strain on the farms' cash resources, but large numbers of farmers have nevertheless been able to find the necessary money. In East Kitutu in

1967, expansion of the dairy industry was being held up by a shortage of heifers, not of money to buy them. In many instances, however, farmers are prevented from making these larger purchases by their inability to borrow the necessary money.

Adoption of new methods has been inhibited by inefficiencies in the marketing of farm tools, agricultural chemicals, seed, and other inputs required by the new technology. In general, private firms have been unwilling to underwrite the costly introduction of new fertilizers and pesticides to farmers unless they could be permitted to enjoy some of the profits of these innovations, once widespread acceptance was assured. Governments, fearing exploitation of farmers, have been reluctant to grant private sellers this degree of monopoly and have instead assumed the distribution responsibility themselves, usually with unsatisfactory consequences. Thinly staffed marketing boards, hard pressed to handle the assembly and transport of grain, experience even greater difficulty in stocking and purveying items like hybrid seeds, chemical fertilizers, and pesticides. Extension agents whose primary tasks are teaching and advising are usually poor merchants.

Increased production for sale and the accompanying specialization in crops and in productive activities are only possible if someone is willing to buy the added output. African farmers have utilized export opportunities for sale to good advantage and undoubtedly will continue to do so. Export markets can be expected to grow, but for many African countries the major sources of market expansion must be sought in domestic demand. The countries of tropical Africa satisfy the greater part of their food requirements with domestic production, and opportunities for import substitution are modest.[3] Intra-African trade in foodstuffs—for example, maize in eastern and southern Africa—offer opportunities to low-cost producers if their realization is not thwarted by nationalist policies. Nevertheless, the principal market for African farmers will be found in supplying the populations of their own countries.

Market demand for foodstuffs will grow with population, personal income, and urbanism. However, the most important stimulus for some decades to come will probably be that resulting from the substitution of purchased food for own consumption. This phenomenon affects both urban and rural populations. City dwellers in many areas still rely to some degree on foodstuffs produced on their own farms or on those of their kinsmen,[4] and rural residents, like the Sukuma of Geita District, Tanzania, increas-

---

[3]Wheat, sugar, and rice are notable exceptions.

[4]In 1960, approximately two-thirds of the residents of Ibadan, Nigeria—then tropical Africa's largest city—were farmers.

ingly rely on the market for their staple food supplies.[5] Not all rural residents are farmers only; most produce nonagricultural goods and services as well. Nor do all farmers grow food crops sufficient to the needs of their own households. These country people obtain at least part of their food requirements from the market or from their neighbors. This specialization in the small already manifests itself in increased market demand for local foodstuffs.

The effective farm demand for food crops can be expected to increase rapidly with growth of income and with the transformation to a market economy. As it does, specialized production will be stimulated; and as specialized production increases, so also will the purchase of foodstuffs. The economic achievements of tropical African agriculture to date were made possible by the opening of world markets to African farmers. This great expansion of trade in "cash crops" is only beginning to be reflected in the marketing of food crops. The task is to encourage the commercialization of crops intended for domestic consumption. It will be a great error, however, if new attention to African food production results in neglect of export-based agriculture. Continued vigor in this generator of agricultural growth is critical for the development of the food sector as well as for that of the economy at large.

Our parallel study of staple food marketing found a considerable range in the efficiency of tropical African marketing systems, although it is generally true that private firms have performed much better than public firms in the distribution of domestically consumed food crops. Parastatal marketing boards have been concerned mostly with the flow of export crops from farm to port and have avoided the more complex food trade. A notable exception is the trade in maize in the former British territories of eastern and southern Africa. The maize boards of English-speaking Africa were originally set up to protect European producers from the competition of African farmers at a time when the principal markets were expected to be overseas or in the mining districts. In every instance, the maize boards have failed in their attempts to monopolize the trade entrusted to them—the extent of their failure being a measure of the inefficiency of their operations as compared with those of private merchants.

Despite the relatively better performance of the private trade, it too suffers from numerous inefficiencies and imperfections resulting from deficiencies of transport, communications, and physical market facilities.

[5]As early as 1950, the Tanganyika Department of Agriculture complained because the Sukuma of Lake Province were relying on the market for supplies of cassava (Jones 1960: 119).

African merchants in general perform well under difficult conditions, but problems of spatial arbitrage and integration are likely to impair their ability to respond to the changes in market supply and demand that lie ahead.

## The Design of Innovations

Our studies did not find elements in the present sociocultural system that might bear directly on the acceptability of particular innovations—which was a bit surprising. Perhaps we did not look hard enough. The societies that were examined differ greatly in social organization and environment. At the very least there might have been some indications that innovation was harder or easier in a task performed by a man, or a woman, or a child. This question remains unanswered.

On the other hand, the existing agricultural technology, as contrasted with the social order, has constrained the range of feasible innovations, the speed with which they can be learned, and the willingness of farmers to undertake them. Thorough understanding of existing farming systems and step-by-step modification of them, therefore, possess great advantages. Single-trait innovations are well adapted to this approach when they are employed in a sequence that permits each to build on changes that have already been achieved. The changes can be technical or economic. Often the adoption of a low-risk annual crop that costs little to grow has enabled farmers to accumulate large enough financial reserves so that they have been willing to undertake the cultivation of more costly perennial crops. The training of oxen for plowing and the purchase of a crude plow or ridger has often been the first step in the acquisition of a complete set of ox-drawn equipment. Sometimes one single-trait innovation made mandatory the adoption of a series, as when the first tea stump was planted. Certainly it was easier for farmers to keep exotic dairy cattle if they already had some knowledge of animal feeding and pest control, if they had already fenced their cropland and put in some sort of water supply to the farm, and if they had acquired the regular work habits that dairying demands.

## Policy Implications

It should not be too surprising that a research team made up of two economists, a social anthropologist, and an agricultural scientist, each with long experience in tropical Africa, should end up a study of agricultural change with the simple conclusion that profitability is the primary determinant of farmer behavior. At any rate, it did not seem surprising when the study was concluded, although it probably would have done at the outset.

If profitability is the prime mover, it is not hard to identify the principal elements at one step removed: the prices of the products, the purchase prices and opportunity costs of resources used in their production, and the physical efficiency of the production process. These elements of production point the direction for action and for research. Prices are affected by the efficiency of the marketing system and also by costs of transporting, storing, and processing. Opportunity costs are affected by all of these, by the degree to which farmers participate in the market economy, and by the ways in which other enterprises in the economy are organized for production. And technical efficiency is a manifestation of the state of the arts—the technical and managerial skills that farmers have developed over the generations—and of the findings of public and private agricultural research agencies.

From the foregoing summary of our findings, it is relatively easy to sketch a broad policy design for an agriculture based on small farms. It would include more money for research; emphasis on simple yield-increasing innovations that do not require a massive educational program in order to achieve widespread adoption; development of better methods for supplying purchased inputs to farmers; and increased national capacity for produce marketing and spatial arbitrage.

When it comes to translating some of these broad objectives into specific policy directives, however, it is necessary to consider another important characteristic of the national economies—the variability among farmers and the economic and physical environments in which they operate. An agriculture made up of many small producing units that are coordinated through the market is designed to accommodate to and take advantage of this variability. On the other hand, it is ill suited to direct control in detail by orders emanating from a national or provincial center. Both the transmission of such directives and the monitoring of their execution are costly and would require a manyfold increase in the present agricultural department staffs. The implication of this conclusion for research is much greater emphasis on the study of the economics of present farm enterprises, to the end that farmers may be provided with a choice of innovations and may themselves work out the best combinations for their own circumstances.

Traditional farming methods—long fallow, a mixture of crops in the field, and cultivation with hoe and cutlass—have been well suited to African conditions. By using these methods, African farmers have grown a wide range of new crops that supply the present export-crop industries. The possibilities of traditional farming are not yet exhausted. Adoption of improved varieties of existing crops and of new crops, better timing of farm operations, improved plant populations, improved crop mixtures in the field, and specialized production can result in substantial increases in productivity without much increase in effort. Ox cultivation, if this can be considered traditional, has potential for permitting more effective use of farm labor. It is also true in tropical Africa, as it probably is in other countries, that if all farmers farmed as well as the best farmers do, great increases in output and productivity would ensue.

But the time is not far off when farming by traditional methods, no matter how improved, will not satisfy the requirements of rapidly growing populations or achieve the increases in returns that constitute agricultural progress. As population has risen, fallow periods have had to be shortened. This intensification in cropping, without intensification of farming methods, has led to declines in output per acre. Major technical changes, primarily the employment of new yield-raising inputs, are required. Agricultural research to adapt to African conditions methods that have been developed elsewhere, and to develop new methods, is therefore of primary importance.

The labor force of tropical Africa is growing faster than it can be employed in urban jobs, and it will almost certainly continue to do so for several decades to come. The pattern of farm development that relies on widespread mechanization of field activities would only exacerbate problems of unemployment. Mechanization has a place in African farming, primarily as it can improve the timeliness of sowing and harvesting and reduce the cost of transport. It will be most beneficial when it employs power equipment that is designed for small units. Mechanization of threshing, drying, milling, and other processing also has promise. It can increase the capacity of a specialized agriculture to expand the supplies going to domestic markets, thus reducing costs to consumers. It will also free rural and urban housewives from laborious pounding and grinding and enable them to spend more time in the care and education of their children. But mechanization will not help with the principal farm problem of declining soil fertility as the old long-fallow system breaks down.

It will be necessary to increase expenditures for agricultural research greatly in the years ahead. The new international agricultural research centers will provide powerful added knowledge that can be tapped by competent national research organizations. The costs of research can also be reduced by cooperation and communication among African states, although there will always remain a need for on-the-spot trials.

Research targets should be set with understanding both of national needs and of the requirements of farmers. New techniques must bring farmers rewards that are greater than their costs, and farmers must perceive them to do so.

Agricultural development is a learning process, and extension education can play an important part. It is of little value, however, if it has nothing to extend. Even when attractive new methods are known, complete reliance on the extension service for their dissemination would be excessively costly in a smallholder agriculture. Extension agents can introduce new methods, but widespread dissemination requires other means. Farmers learn from advertising and from salesmen as well as from extension programs. Yet most of what they know about farming they learn from one another. Innovations that lead to easily visible results are likely to spread throughout a farming community without much need for extension agents. In this regard, there is also great advantage in making the new inputs available in very small amounts—a few pounds of a new seed, for example, rather than requiring a minimum purchase of a hundredweight—for farmers who want to experiment without too much risk. Because of the variability in local conditions, the research staff should be on the alert for reasons why farmers have failed to adopt their recommendations.

Few innovations can be perfected in the laboratory or on the experimental farm. It is not possible to be sure of the suitability of a new practice or product until it has been tested by ordinary farmers in their own fields. Frequently a cycle of repeated farm testing and modification may be required. Widespread release or advocacy of a practice that is not reliable can have most unfortunate results, not the least of which is its effect on the willingness of farmers to try new methods.

Willingness to change customary practices implies a belief that things can be better, and these expectations are determined for most men, excepting the perennial optimists, by their experience with previous innovations. There can be no doubt that the attitude toward change is modified by experience, and there are probably more illustrations in tropical Africa of changes that have made farmers more conservative than of ones that have made them more willing to experiment.

We stated at the outset our conviction that African farmers are motivated by various desires and concerns, and that economic drive varies from individual to individual. People also vary in the kind and quality of the skills with which they undertake to satisfy their desires, and in their curiosity ingenuity, and innovativeness. An egalitarianism that fails to recognize these variations in ambition and abilities wastes a most valuable human resource.

Farmers need freedom to try out new ways and to invent new ways, as contrasted with countrywide prescription or imposition, say, of fertilizer application rates, or the requirement of tied purchases of seed and fertilizer. One of the advantages of single-trait innovations is that farmers can more easily experiment with them.

## Small Farming Systems

Burgeoning export-crop industries in tropical Africa have demonstrated that productive agriculture can grow upon the basis of small farms, mostly of less than thirty acres and typically with one or two cropped acres for each member of the farm family. The experience in various parts of tropical Africa demonstrates further that such an agriculture can be responsive to changing markets and changing technical knowledge, and that it can achieve sustained economic growth if a set of realistic conditions is fulfilled. It would be incorrect to characterize all the agriculture of any single country of tropical Africa today as efficient, responsive, and productive. But it is appropriate to so describe the agriculture of particular areas—for example, East Kitutu in Kenya and Mazabuka in Zambia. Other areas, like Teso and Akim-Abuakwa, enjoyed this status in earlier years. Given the political will, there would seem to be no insuperable economic, cultural, or technical obstacles to extending efficient farming to a national scale in most African countries.

## Agriculture in African Development Strategies

It is clear that accelerating the increase in agricultural production will not in itself permit the tropical African countries to achieve their national development goals. Development involves transformation of the predominantly rural structure of African economies and the emergence of new institutions, values, attitudes, and behavior.

The progressive modernization of a country's rural households will affect attitudes and behavior more favorably than relying primarily on large mechanized farming units that more or less bypass most farm households. Widely shared increases in farm incomes, resulting from progressive modernization, generate increased demand for inexpensive mass-produced goods that can be provided by local manufacturers. This kind of agricultural development is less likely to be constrained by shortages of capital and foreign exchange; and by strengthening and diffusing entrepreneurial and technical competence among a country's manufacturing firms, it facilitates the production of many other products and contributes to the development of capital goods industries.

There is evidence, too, that development strategy which leads to broad participation in opportunities for technical and economic advance and rising aspirations among the rural population helps to create an environment favorable to the success of family planning programs. Moreover, the prospects for securing the financial and political support required to implement effective rural health programs are likely to be considerably better under these circumstances.

Even under colonial rule, tropical Africa was relatively free from the kinds of dualistic agricultural systems that characterize many other parts of the world. It seems certain that the people of Africa will be best served if this can continue to be true.

# Sources Cited

Abercrombie, K. C. "Subsistence Production and Economic Development." *Monthly Bulletin of Agricultural Economics and Statistics* 14 (1965):1-8.

Adamu, S. O. "Expenditure Elasticities of Demand for Household Consumer Goods." *Nigerian Journal of Economic and Social Studies* 8 (1966):481-90.

Allan, William. *The African Husbandman.* Edinburgh, 1965.

Alvis, Q., and Temu, P. E. "Marketing Selected Staple Foodstuffs in Kenya." West Virginia University, Department of Agricultural Economics and Office of International Programs, 1P-25. Morgantown, March 1968.

Amin, Samir. *"La politique coloniale française a l'égard de la bourgeoisie commerçante senegalaise (1820-1960)."* In *The Development of Indigenous Trade and Markets in West Africa,* ed. Claude Meillassoux. Oxford, 1971.

Andrews, D. J. "Responses of Sorghum Varieties to Intercropping." *Experimental Agriculture* 10 (1974):57-63.

Angladette, André, and Deschamps, Louis. *Problèmes et perspectives de l'agriculture dans les pays tropicaux.* Paris, 1974.

Anthony, K. R. M., and Johnston, B. F. "Field Study of Agricultural Change: Northern Katsina, Nigeria." Food Research Institute Study of Economic, Cultural, and Technical Determinants of Agricultural Change in Tropical Africa, Preliminary Report No. 6. Stanford, 1968, processed.

Anthony, K. R. M., and Uchendu, V. C. "Agricultural Change in Mazabuka District, Zambia." *Food Research Institute Studies* 9 (1970):215-67.

––––––. *Agricultural Change in Geita District, Tanzania.* Nairobi, 1974.

Apthorpe, R. J., ed. *From Tribal Rule to Modern Government.* Lusaka, 1959.

––––––. "The Introduction of Bureaucracy into African Politics." *Journal of African Administration* 12 (1960):125-34.

Arnold, M. H. "Cotton Improvement in East Africa." In Leakey, 1970.

Arnon, I. *Organisation and Administration of Agricultural Research.* London, 1968.

Bartlett, C. D. S. "The Spread of Rain-Grown Cotton in Tropical Africa." Ph.D. dissertation, Stanford University, 1973.

Bateman, M. J. "Aggregate and Regional Supply Functions of Ghanaian Cocoa, 1946-1962." *Journal of Farm Economics* 47 (1965):384-401.

––––––. "Supply Relations for Perennial Crops in the Less-Developed Areas." In Wharton, 1969.

Bates, R. H. *Rural Responses to Industrialization: A Study of Village Zambia.* New Haven, 1976.

Bauer, P. T. *West African Trade.* Cambridge, 1954.

Bayagagaire, J. M. "A Summary of Agriculture in Kigezi District, Western Region, Uganda." Kigezi, 1962, processed.

Beckett, W. H. *Akokoaso: A Survey of a Gold Coast Village*. London, 1944.

Belshaw, D. G. R. *Crop Production Data in Uganda: A Statistical Evaluation of International Agricultural Census Methodology*. Overseas Development Group, Development Studies Discussion Paper No. 7, East Anglia University. Norwich, 1975.

Belshaw, Deryke, and Chambers, Robert. "A Management Systems Approach to Rural Development." University of Nairobi, Institute for Development Studies Discussion Paper No. 161. Nairobi, January 1973.

Belshaw, D. G. R., and Hall, Malcolm, "The Analysis and Use of Agricultural Experimental Data in Tropical Africa." *Journal of Rural Development* 5 (1972): 39-71.

Benneh, George. "Small-Scale Farming Systems in Ghana." *Africa* 43 (1973):134-46.

Bennett, M. K. "An Agroclimatic Mapping of Africa." *Food Research Institute Studies* 3 (1962):195-216.

Berg, Elliot. "The Recent Economic Evolution of the Sahel." Center for Research on Economic Development, University of Michigan. Ann Arbor, June 1, 1975.

Bernard, F. E. *East of Mount Kenya: Meru Agriculture in Transition*. Munich, 1972.

Berry, S. S. *Cocoa, Custom, and Socio-Economic Change in Rural Western Nigeria*. Oxford, 1975.

Bohannan, P. J., and Dalton, George, eds. *Markets in Africa*. Evanston, Ill., 1962.

Boserup, Ester. *The Conditions of Agricultural Growth: The Economics of Agrarian Change under Population Pressure*. Chicago, 1965.

Boshoff, W. H. "Farming Systems in African Mechanization of Agriculture." In NAS, 1968.

Bottrall, A. F. "Financing Small Farmers: A Range of Strategies." In Hunter et al., 1976.

Bovill, E. W. *Golden Trade of the Moors*. London, 1958.

Boyce, J. K., and Evenson, R. E. *National and International Agricultural Research and Extension Programs,* Agricultural Development Council. New York, 1975.

Brown, L. H. "Development and Farm Planning in the African Areas of Kenya." *East African Agricultural Journal* 23 (1957):67-73.

———. "Agricultural Change in Kenya: 1945-1960." *Food Research Institute Studies* 8 (1968):35-90.

Brown, W. T., Evans-Jones, P., and Innes, D. "Implements." In Jameson et al., 1970.

Bruel, G. *"Les Cultures indigènes à développer en A.E.F.: Le Tabac et le manioc Batéké: L'Huile de palme: Les Fermes-écoles à créer."* In *Congrès d'Agriculture, 21-25 Mai 1918, Comptes Rendu des Travaux* IV. Paris, 1920.

Bunting, A. H. ed. *Change in Agriculture*. London, 1970.

Byerlee, Derek, and Eicher, C. K. "Rural Employment and Economic Development: Theoretical Issues and Empirical Evidence from Africa." In *Agricultural Policy in Developing Countries: Proceedings of a Conference held by the Inter-*

*national Economic Association at Bad Godesberg, West Germany,* ed. Nurul Islam. London, 1974.

Byerlee, Derek, Eicher, C. K., Liedholm, Carl, and Spencer, D. S. C. "Rural Employment in Tropical Africa: Summary of Findings." Njala University College and Michigan State University, African Rural Economy Program, Working Paper No. 20. February 1977.

Caldwell, J. C. *The Sahelian Drought and Its Demographic Implications.* Overseas Liaison Committee Paper No. 8. Washington, December 1975.

Caves, R. E. " 'Vent for Surplus' Models of Trade and Growth." In *Trade, Growth and the Balance of Payments: Essays in Honor of Gottfried Haberler.* R. E. Baldwin et al. Chicago, 1965.

Chambers, Robert, and Belshaw, Deryke. "Managing Rural Development: Lessons and Methods from Eastern Africa." University of Sussex, Institute of Development Studies Discussion Paper No. 15. Brighton, June 1973.

Chilalo Agricultural Development Unit (CADU). *Project Description.* Addis Ababa, October 1971.

Cleave, J. H. *African Farmers: Labor Use in the Development of Smallholder Agriculture.* New York, 1974.

Collinson, M. P. *Farm Management in Peasant Agriculture.* New York, 1972.

Colson, Elisabeth. "Trade and Wealth Among the Tonga." In Bohannan and Dalton, 1962.

Cooper, St. G. C. *Agricultural Research in Tropical Africa.* Nairobi, 1970.

Côte d'Ivoire. *Service de la Statistique, Enquête nutrition—Niveau de vie: Subdivision de Bongouanou 1955–1956.* Paris, 1958.

Cramer, H. H. *Plant Protection and World Crop Production.* Leverkusen, 1967.

Crossley, D. P., and Kilgour, J. J. "The Snail in Malawi: Testing a Novel Cultivation Device." *World Crops* 26 (1974):170–72.

Curtis, D. L., et al. Botany Section in *Annual Report,* Institute for Agricultural Research. Samaru, 1963.

Darrow, Ken, and Pan, Rick. *Appropriate Technology Sourcebook.* 2d ed. Stanford, 1976.

Davis, L. E., and North, D. C. *Institutional Change and American Economic Growth.* Cambridge, Mass., 1971.

Dean, E. R. "Economic Analysis and African Response to Price." *Journal of Farm Economics* 47 (1965):402–9.

de Smet, R. E. *"Une Enquête budget-temps dans les Uélé République du Zaïre."* In *Etudes de geographie tropicale offertes à Pierre Gourou.* Paris, 1972.

Devred, R. "Agricultural Research Programmes—Ecological Bases—Basic Principles and General Measures for Strengthening Co-operation." In NAS, 1968.

de Wilde, J. C., et al. *Experiences with Agricultural Development in Tropical Africa.* I, *The Synthesis.* II, *The Case Studies.* Baltimore, 1967.

Doggett, H. "The Application of Modern Plant Breeding Methods to Mainly Self-Pollinated Crops." *African Soils* 15 (1970a):629–42.

―――. "Sorghum Improvement in East Africa." In Leakey, 1970b.

―――. *Sorghum.* London, 1970c.

Doyle, C. J. "Productivity, Technical Change, and the Peasant Producer: A Profile of the African Cultivator." *Food Research Institute Studies* 13 (1974):61–76.

Drachoussoff, V. "Agricultural Change in the Belgian Congo: 1945-1960." *Food Research Institute Studies* 5 (1965):137-201.

Dumont, René. *Types of Rural Economy*. London, 1957.

East Africa Royal Commission, 1953-1955. *Report*. London, 1955.

Ehrlich, Cyril. "Building and Caretaking: Economic Policy in British Tropical Africa, 1890-1960." *Economic History Review,* 2d Series, 24 (1973): 649-67.

Elkan, Walter. "Africana: Concepts in the Description of African Economics." *Journal of Modern African Studies* 14 (1976):691-95.

Etherington, D. M. "An Econometric Analysis of Smallholder Tea Growing in Kenya." Ph.D. dissertation, Stanford University, 1970.

Fallers, L. A. *Bantu Bureaucracy*. Cambridge, 1956.

_____. "Are African Cultivators to Be Called 'Peasants'?" *Current Anthropology* 2 (1961):108-10.

Farbrother, H. G. *Crop Physiology in Cotton Research Reports, Republic of the Sudan, 1970-71*. Cotton Research Corporation. London, 1973.

Food and Agriculture Organization of the United Nations (FAO). *Agricultural Development in Nigeria 1965-1980*. Rome, 1966.

_____. *Indicative World Plan for Agricultural Development to 1975 and 1985*. Vol. I, Provisional Regional Study No. 3. Rome, 1968a.

_____. *Indicative World Plan for Agricultural Development to 1975 and 1985*. Vol. II, Provisional Regional Study No. 3. Rome, 1968b.

_____. *Trade Yearbook 1970*. Rome, 1971a.

_____. *Conference on the Establishment of Cooperative Agricultural Research Programmes between Countries with Similar Ecological Conditions: Guinean Zone, Africa*. Rome, 1971b.

_____. *Production Yearbook 1975*. Rome, 1976a.

_____. *Trade Yearbook 1975*. Rome, 1976b.

_____. *Commodity Review and Outlook 1975-76*. Rome, 1976c.

_____. *State of Food and Agriculture 1975*. Rome, 1976d.

_____. *Annual Fertilizer Review 1975*. Rome, 1976e.

_____. "Special Feature: FAO Indices of Agricultural Production: Agriculture, Food, Crops, Cereals and Livestock Products." *Monthly Bulletin of Agricultural Economics and Statistics* 26 (1977):20-29.

Forbes, Frederick E. *Dahomey and the Dahomans: Being the Journals of Two Missions to the King of Dahomey, and Residence at His Capital, in the Years 1849 and 1950,* vol. I. London, 1851.

Forde, Daryll. *Habitat, Economy and Society: A Geographical Introduction to Ethnology*. 8th ed. London, 1950.

Fuggles-Couchman, N. R. "Agricultural Change in Tanganyika:1945-1960." *Food Research Institute Studies,* Supplement to 5 (1965): 4-98.

Galletti, R., Baldwin, K. D. S., and Dina, I. O. *Nigerian Cocoa Farmers*. London, 1956.

Gann, L. H. *A History of Northern Rhodesia: Early Days to 1953*. London, 1964.

Gascon, J. P. "Oil Palm Selection by IRHO." In NAS, 1968.

Gerhart, John. *The Diffusion of Hybrid Maize in Western Kenya*. Abridged by CIMMYT. Mexico City, 1975.

Ghana, Office of Government Statistics. *Survey of Population and Budgets of Cocoa Producing Families in Oda-Swedru-Asamankese Area: 1955–56.* Statistics and Economic Papers No. 6. Gold Coast, 1958.

Gilbert, E. H. "The Marketing of Staple Foods in Northern Nigeria." Ph.D. dissertation, Stanford University, 1969.

Goddard, A. D. "Industry." In *Zaria and Its Region: A Nigerian Savanna City and Its Environs,* ed. M. J. Mortimore. Zaria, 1970.

Goody, Jack. *Technology, Tradition, and the State in Africa.* London, 1971.

Gouvernement Général de l'Afrique Occidentale Française, Inspection Générale de l'Agriculture. *Le Centre de Recherches Agronomiques de Bambey au Service de la Production.* Dakar, 1956.

Gray, Richard, and Birmingham, David, eds. *Pre-Colonial African Trade: Essays on Trade in Central and Eastern Africa before 1900.* London, 1970.

Green, R. H., and Hymer, S. H. "Cocoa in the Gold Coast: A Study in the Relations between African Farmers and Agricultural Exports." *Journal of Economic History* 26 (1966):299–319.

Gulliver, P. H. "The Evolution of Arusha Trade." In Bohannan and Dalton, 1962.

Hägerstrand, Torsten. "Diffusion of Innovation." In *International Encyclopedia of the Social Sciences,* ed. D. L. Sills, vol. IV. New York, 1968.

Hailey, Lord. *An African Survey.* London, 1938.

Hance, W. A. *Population, Migration, and Urbanization in Africa.* New York, 1970.

Harrison, M. N. "Maize Improvement in East Africa." In Leakey, 1970.

Hay, M. J. "Economic Change in Luoland: Kowe, 1890–1945." Ph.D. dissertation, University of Wisconsin, 1972.

Helleiner, G. K. *International Trade and Economic Development.* Middlesex, 1972.

Heyer, Judith. "Some Problems in the Valuation of Subsistence Output." University of Nairobi, Institute for Development Studies Discussion Paper No. 14. Nairobi, August 1965.

―――. "Achievements, Problems and Prospects in the Agricultural Sector." In Heyer et al., 1976.

Heyer, Judith, Maitha, J. K., and Senga, W. M., eds. *Agricultural Development in Kenya: An Economic Assessment.* Nairobi, 1976.

Hildebrand, P. E. *Generando tecnologia para agricultores tradicionales: Una metadologia multidisciplinaria (Generating technology for traditional farmers: A multidisciplinary methodology).* Socioeconomia Rural, Instituto de Ciencia y Tecnologia Agricolas, Sector Publico Agricola. Guatemala, December 1976. In English and Spanish.

Hill, Polly. *The Gold Coast Cocoa Farmer: A Preliminary Survey.* London, 1956.

―――. *The Migrant Cocoa Farmers of Southern Ghana.* Cambridge, 1963.

―――. "The Myth of the Amorphous Peasantry: A Northern Nigerian Case Study." *Nigerian Journal of Economic and Social Affairs* 10 (1968):239–260.

―――. *Studies in Rural Capitalism in West Africa.* Cambridge, 1970.

Hogendorn, J. S. "The Vent-for-Surplus Model and African Cash Agriculture to 1914." *Savanna* 5 (1976):15–28.

Hooker, A. L. "Crop Vulnerability and Breeding for Disease Resistance." *Plant Protection Bulletin* 20 (1972):18–20.

Hopkins, A. G. *An Economic History of West Africa*. New York, 1973.

Houthakker, H. S. "An International Comparison of Household Expenditure Patterns, Commemorating the Centenary of Engel's Law." *Econometrica* 25 (1957):532–51.

Howard, H. W., et al. "Problems in Breeding for Resistance to Disease and Pests, Part II." In *Annual Report, Plant Breeding Institute*. Cambridge, 1969.

Hunter, Guy. "Agricultural Administration and Institutions." *Food Research Institute Studies* 12 (1973):233–51.

Hunter, Guy, Bunting, A. H., and Bottrall, Anthony, eds. *Policy and Practise in Rural Development: Proceedings of the Second International Seminar on Change in Agriculture, Reading 9–19 September 1974*. New York, 1976.

Hymer, Stephen, and Resnick, Stephen. "A Model of an Agrarian Economy with Nonagricultural Activities." *American Economic Review* 59, Part 1 (1969): 493–506.

Idris, Hussein. "The Evolution of Government Agricultural Research in the Sudan." *Sudan Agricultural Journal* 4 (1969):1–12.

_____. Forword to *Cotton Growth in the Gezira Environment*. Cambridge, 1970.

International Bank for Reconstruction and Development (IBRD). *World Tables 1976*. Washington, 1976.

International Labour Office (ILO). *Employment, Incomes, and Equality: A Strategy for Increasing Productive Employment in Kenya*. Geneva, 1972.

Jamal, Vali. "The Role of Cotton and Coffee in Uganda's Economic Development." Ph.D. dissertation, Stanford University, June 1976.

Jameson, J. D., et al. *Agriculture in Uganda*. London, 1970.

Johnson, R. W. M. "African Agricultural Development in Southern Rhodesia: 1945–1960." *Food Research Institute Studies* 4 (1964):165–223.

Johnston, B. F. "Changes in Agricultural Productivity." In *Economic Transition in Africa*, ed. M. J. Herskovits and Mitchell Harwitz. Evanston, 1964.

Johnston, B. F., and Kilby, Peter. *Agriculture and Structural Transformation: Economic Strategies in Late-Developing Countries*. New York, 1975.

Jones, B. K. "The Beginning of the Iron Age in Subsaharan Africa." Unpublished working paper. Stanford, 1971.

Jones, E. "Nutrient Cycle and Soil Fertility on Red Ferralitic Soils." In *Transactions of International Congress of Soil Science*, vol. III. Adelaide, 1968b.

_____. "Crop Husbandry." In *Progress Reports from Experiment Stations Uganda*, Cotton Research Corporation. London, 1971.

Jones, G. E. "The Diffusion of Agricultural Innovations." *Journal of Agricultural Economics* 15 (1963):387–405.

_____. "The Adoption and Diffusion of Agricultural Practices." *World Agricultural Economics and Rural Sociology Abstracts* 9 (1967):1–34.

Jones, W. O. "Manioc: An Example of Innovation in African Economies." *Economic Development and Cultural Change* 5 (1957): 99–117.

_____. *Manioc in Africa*. Stanford, 1959.

_____. "Economic Man in Africa." *Food Research Institute Studies* 1 (1960): 107–34.

_____. "Environment, Technical Knowledge, and Economic Development in Tropical Africa." *Food Research Institute Studies* 5 (1965):101–16.

_____. "Labor and Leisure in Traditional African Societies." *Items* 22 (1968a): 1–6.

_____. "The Demand for Food, Leisure, and Economic Surpluses." In Wharton, 1969.

_____. "Measuring the Effectiveness of Agricultural Marketing in Contributing to Economic Development: Some African Examples." *Food Research Institute Studies* 9 (1970):175–96.

_____. *Marketing Staple Food Crops in Tropical Africa.* Ithaca, 1972.

Jones, W. O., and Merat, Christian. "Consumption of Exotic Consumer Goods as an Indicator of Economic Achievement in Ten Countries of Tropical Africa." *Food Research Institute Studies* 3 (1962):35–60.

Kaneda, Hiromitsu,. and Johnston, B. F. "Urban Food Expenditure Patterns in Tropical Africa." *Food Research Institute Studies* 2 (1961):229–75.

Katz, E. "The Characteristics of Innovations and the Concept of Compatibility." In *Rehovoth Conference on Comprehensive Planning of Agriculture in Developing Countries.* Rehovoth, August 19–29, 1963.

Kenya, Government of. *Development Plan, 1974–1978.* Nairobi, 1974.

_____, Maize Board Commission of Inquiry. *Report.* Nairobi, 1966.

_____, Ministry of Economic Planning and Development, Statistical Division. *Economic Survey of Central Province 1963/64.* Nairobi, 1968.

Kenyatta, Jomo. *Facing Mount Kenya: The Tribal Life of the Gikuyu.* New York, 1962.

Kettlewell, R. W. "Agricultural Change in Nyasaland: 1945–1960." *Food Research Institute Studies* 5 (1965):229–85.

Keyfitz, Nathan. "The Interlocking of Social and Economic Factors in Asian Development." *Canadian Journal of Economics and Political Science* 25 (1959):34–46.

Kilby, Peter. "The Nigerian Palm Oil Industry." *Food Research Institute Studies* 7 (1967):177–203.

_____. *Industrialization in an Open Economy: Nigeria 1945–1966.* Cambridge, 1969.

King, R. P. "An Analysis of Rural Consumption Patterns in Sierra Leone and Their Employment and Growth Effects." Njala University College and Michigan State University African Rural Economy Program, Working Paper No. 21. East Lansing, March 1977.

Kmietowicz, T., and Webley, P. "Statistical Analysis of Income Distribution in the Central Province of Kenya." *Eastern Africa Economic Review* 7 (1975): 1–26.

Kuznets, Simon. *Modern Economic Growth: Rate, Structure, and Spread.* New Haven, 1966.

Labrousse, G. "Research (in the Broadest Sense) in the Field of Tropical Farm Mechanization (Francophone Countries)." In NAS, 1968.

Lawrance, J. C. D. *Report of the Mission on Land Consolidation and Registration in Kenya, 1965–66.* Nairobi, 1967.

Leakey, C. L. A., ed. *Crop Improvement in East Africa.* Farnham Royal, England, 1970.

Leakey, C. L. A., and Wills, J. B., eds. *Food Crops of the Lowland Tropics.* Oxford, 1977.

Lele, Uma. *The Design of Rural Development: Lessons from Africa*. Baltimore, 1975.

──────. "Designing Rural Development Programmes: Lessons from Past Experience in Africa." In Hunter et al., 1976.

Leonard, D. K. Introduction to *Rural Administration in Kenya*, ed. D. K. Leonard. Nairobi, 1973.

Leurquin, P. P. "Agricultural Change in Ruanda-Urundi: 1945–60." *Food Research Institute Studies* 4 (1963):39–89.

LeVine, R. A. "Africa." In *Psychological Anthropology*, ed. F. L. K. Hsu. Homewood, Ill., 1961.

──────. *Dreams and Deeds: Achievement Motivation in Nigeria*. Chicago, 1966a.

LeVine, R. A., and LeVine, B. B. *Nyansongo: A Gusii Community in Kenya*. New York, 1966b.

Lewis, W. A. *Reflections on Nigeria's Economic Growth*. Paris, 1967.

Lombard, S. C., and Tweedie, A. H. C. *Agriculture in Zambia since Independence*. Lusaka, 1974.

Long, Norman. *Social Change and the Individual: A Study of the Social and Religious Responses to Innovation in a Zambian Rural Community*. Manchester, 1968.

Lynam, J. K. "Palm Oil Production in West Africa: Its Role in World Palm Oil Trade." *Foreign Agriculture* 10 (1972):8–10.

──────. "An Analysis of Technical Change Under Risk in Semi-Arid Agriculture: A Case Study of Smallholders in Kenya." Draft Ph.D. dissertation, Stanford University, 1977.

Lynn, C. W. *Agriculture in North Mamprusi*. Gold Coast, Department of Agriculture Bulletin No. 34. Accra, 1937.

Mabogunje, A. K. *Regional Mobility and Resource Development in West Africa*. Montreal, 1972.

Mackenzie, William. "The Use of Aggregate Data in the Study of Agricultural Change: A Case Study of Uganda, 1955–67." *Food Research Institute Studies* 10 (1971):21–54.

Maina, J. W., and MacArthur, J. D. "Land Settlement in Kenya." In Bunting, 1970.

Makings, S. M. "Agricultural Change in Northern Rhodesia/Zambia: 1945–65." *Food Research Institute Studies* 6 (1966):195–247.

──────. *Agricultural Problems of Developing Countries in Africa*. Oxford, 1967.

Manig, Winifred. "Marketing of Selected Agricultural Commodities in the Baco Area, Ethiopia." Cornell University, Department of Agricultural Economics, Employment and Income Distribution Project, Occasional Paper No. 66. Ithaca, November 1973.

Masefield, G. B. "Agricultural Change in Uganda: 1945–60." *Food Research Institute Studies* 3 (1962):87–124.

──────. *A History of the Colonial Agricultural Service*. Oxford, 1972.

Massell, B. F., and Heyer, Judith. "Household Expenditure in Nairobi: A Statistical Analysis of Consumer Behavior." *Economic Development and Cultural Change* 17 (1969):212–34.

Massell, B. F., and Johnson, R. W. M. "Economics of Smallholder Farming in Rhodesia: A Cross-Section Analysis of Two Areas." *Food Research Institute Studies*, Supplement to 8 (1968):1–74.

Massell, B. F., and Parnes, Andrew. "Estimation of Expenditure Elasticity from a Sample of Rural Households in Uganda." *Bulletin of the Oxford University Institute of Economics and Statistics* 31 (1969):313–29.

Meillassoux, Claude. *Anthropologie économique des Gouro de Côte d'Ivoire: De l'Economie de subsistance à l'agriculture commerciale.* Paris, 1964.

————, ed. *The Development of Indigenous Trade and Markets in West Africa: Studies Presented and Discussed at the Tenth International African Seminar at Fourah Bay College, Freetown, December 1969.* London, 1971.

————, Introduction to Meillassoux, 1971.

Mettrick, H. "Mechanization of Peasant Agriculture in East Africa." In Bunting, 1970.

Mill, J. S. *Principles of Political Economy.* Abridged by J. L. Laughlin. New York, 1902.

Minto, S. D., and Westley, S. B. *Low Cost Rural Equipment Suitable for Manufacture in East Africa.* Nairobi, June 1975.

Miracle, M. P. *Maize in Tropical Africa.* Madison, 1966.

Moerman, Michael. *Agricultural Change and Peasant Choice in a Thai Village.* Berkeley, 1968.

Molster, H. C. "An Attempt to Explain the Variations in Gross Cash Farm Income Between Survey Areas." Food Research Institute E.C.T. Project, Working Paper No. 5. Stanford, 1969.

Monnier, J. "Farm Mechanization in Senegal and Its Effects on Production and Employment." In *Report on the Meeting of the FAO/OECD Expert Panel on the Effects of Farm Mechanization on Production and Employment,* FAO. Rome, 1975.

Montgomery, J. D., and Marglin, S. A. "Measuring a Government's 'Will to Develop.'" In *Policies for Promoting Agricultural Development,* Massachusetts Institute of Technology. Cambridge, Mass., 1965.

Morgan, W. B. "Peasant Agriculture in Tropical Africa." In *Environment and Land Use in Africa,* eds. M. F. Thomas and G. W. Whittington. London, 1969.

Mosher, A. T. *Getting Agriculture Moving: Essentials for Development and Modernization.* New York, 1966.

Muckle, T. B., Crossley, C. P., and Kilgour, J. "The 'Snail': A Low Cost Cultivation System for Developing Countries." *World Crops* 25 (1973):226–28.

Mutti, R. J., and Atere-Roberts, D. N. "Marketing Staple Food Crops in Sierra Leone." University of Illinois, March 1968.

Myint, Hla. "The 'Classical Theory' of International Trade and the Underdeveloped Countries." In *Economic Theory and the Underdeveloped Countries,* by Hla Myint. London, 1971.

National Academy of Sciences (NAS). *Agricultural Research Priorities for Economic Development in Tropical Africa: The Abidjan Conference.* 3 vols. Washington, 1968.

————. *African Agricultural Research Capabilities.* Washington, 1974.

Nekby, B. "Minimum Package Projects in Ethiopia: A Brief Position Note." IBRD, Agricultural Projects Department. Washington, September 10, 1971, processed.

Newbury, C. W. "Prices and Profitability in Early Nineteenth-Century Trade." In Meillassoux, 1971.

Nigeria, National Office of Statistics. *Rural Economic Survey: Rural Consumption Enquiry, Food Items, 1963/64.* RES/1966/3. Lagos, 1966.

Norman, D. W. "Economic Analysis of Agricultural Production and Labor Utilization among the Hausa in the North of Nigeria." Michigan State University African Rural Employment Study Paper No. 4. Ann Arbor, January 1973.

_____. *Inter-Disciplinary Research on Rural Development: The Experience of the Rural Economy Research Unit in Northern Nigeria.* Overseas Liaison Committee Paper No. 6, American Council on Education. Washington, D.C., April 1974.

Nyanteng, V. K. "Economic Organization of Yam Marketing in Ghana." M.S. thesis, University of Ghana, 1969.

Nye, P. H., and Greenland, D. J. *The Soil Under Shifting Cultivation.* Farnham Royal, England, 1960.

Ogada, F., and Allan, A. Y. "Maize Research in Kenya." *Span* 11 (1968): 143-49.

Oguntoye, O. A. "Occupational Survey of Old Bussa." NISER. Ibadan, October 1968, processed.

Okali, C., and Kotey, R. A. "Akokoaso: A Resurvey." ISSER, University of Ghana, Technical Publication 15. Legon, 1971.

Okigbo, B. N. "Fitting Research to Farming Systems." In Hunter et al., 1976.

Okoth-Ogendo, H. W. O. "African Land Tenure Reform." In Heyer et al., 1976.

Okurume, Godwin. *The Food Crop Economy in Nigerian Agricultural Policy.* Michigan State University, CSNRD-31. East Lansing, February 1969.

Oluwasanmi, H. A., et al. "Uboma: A Socio-economic and Nutritional Survey of a Rural Community in Eastern Nigeria." World Land Use Survey, Occasional Paper 6. Ebbinford, Bude, England, 1966.

Onakomaiya, S. O. "The Spatial Structure of Internal Trade in Delicacy Foodstuffs in Nigeria." Ph.D. dissertation, University of Wisconsin, 1970.

Onyemelukwe, J. O. C. "Staple Food Trade in Onitsha Market: An Example of Urban Market Distribution Function." Ph.D. dissertation, University of Ibadan, 1970.

Ord, H. W., et al. *Projected Level of Demand, Supply, and Imports of Agricultural Products in 1965, 1970, and 1975.* USDA, ERS/FAS and University of Edinburgh, March 1964.

Park, Hardie. "Estimating Monetary Demand for Agricultural Products in Tanganyika and Nigeria." Working Paper No. 1, Food Research Institute. Stanford, June 1969.

Parkin, David, ed. *Town and Country in Central and Eastern Africa: Studies Presented and Discussed at the Twelfth International African Seminar, Lusaka, September 1972.* London, 1975.

Pearson, S. R., and Cownie, John, eds. *Commodity Exports and African Economic Development.* Lexington, Mass., 1974.

Peat, J. E., and Brown, K. J. "The Yield Responses of Rain-Grown Cotton at Ukiriguru in the Lake Province of Tanganyika." *Empire Journal of Experimental Agriculture* 30 (1962):215-31.

Pehaut, Yves. "*L'arachide au Niger.*" In *Etudes d'économie africaine,* by Yves Pehaut and J. M. Fonseqrive. Paris, 1970.

Pélissier, Paul. *Les Paysans du Sénégal: Les civilisations agraires du Cayor à la Casamance*. Saint-Yrieix, Haute-Vienne, 1966.

Phillips, John. *Agriculture and Ecology in Africa: A Study of Actual and Potential Development South of the Sahara*. London, 1959.

Phillipson, D. W. "The Chronology of the Iron Age in Bantu Africa." *Journal of African History* 14 (1975):321–42.

Polanyi, Karl. "The Economy as Instituted Process." In *Trade and Market in the Early Empires: Economies in History and Theory,* eds. Karl Polanyi, C. M. Arensberg, and H. W. Pearson. Glencoe, Ill., 1957.

Poleman, T. T. "The Food Economies of Urban Middle Africa: The Case of Ghana." *Food Research Institute Studies* 2 (1961):121–74.

Pudsey, D. "The Economics of Outgrower Tea Production in Toro, Western Uganda." Uganda, Ministry of Agriculture and Co-operatives. Kampala, 1967.

Raynaud, Edgar. "The Time Concept in the Evaluation of Rural Underemployment and Leisure Time Activities." *Social Science Information* 8 (1969):59–86.

Redfield, Robert. *Peasant Society and Culture: An Anthropological Approach to Civilization*. Chicago, 1956.

Reining, C. C. *The Zande Scheme: An Anthropological Case Study of Economic Development in Africa*. Evanston, Ill., 1966.

Reusse, Eberhard. *Ghana's Food Industries 1968*. FAO, United Nations Development Program. Rome, April 1968.

Roberts, Andrew. "Nyamwezi Trade." In *Pre-Colonial African Trade,* eds. Richard Gray and David Birmingham. London, 1970.

Rogers, E. M. "Motivations, Values, and Attitudes of Subsistence Farmers: Toward a Subculture of the Peasantry." In Wharton, 1969.

Rosenberg, Nathan. *Perspectives on Technology*. Cambridge, Mass., 1976.

Russell, E. W. "Climate and Crop Yields in the Tropics." *Cotton Growing Review* 44 (1967):87–99.

Ruthenberg, Hans. *Farming Systems in the Tropics*. 2d ed. Oxford, 1976.

Saylor, R. G. *The Economic System of Sierra Leone*. Durham, N.C., 1967.

Schönherr, S., and Mbugua, E. S. "New Extension Methods to Speed Up Diffusion of Agricultural Innovations." University of Nairobi, Institute for Development Studies Discussion Paper No. 200. Nairobi, 1974.

Schultz, T. W. *Transforming Traditional Agriculture*. New Haven, 1964.

Senga, W. M. "Kenya's Agricultural Sector." In Heyer et al., 1976.

Shapiro, K. A. "Efficiency and Modernization in African Agriculture: A Case Study in Geita District, Tanzania." Ph.D. dissertation, Stanford University, 1973.

Smith, Adam. *The Wealth of Nations*. New York, 1937.

Smock, D. R. "Cultural and Attitudinal Factors Affecting Agricultural Development in Eastern Nigeria." *Economic Development and Cultural Change* 18, Part 1 (1969):110–24.

Snyder, D. W. "An Econometric Analysis of Household Consumption and Saving in Sierra Leone." Ph.D. dissertation, Pennsylvania State University, 1971.

Stevens, R. D. *Elasticity of Food Consumption Associated with Changes in Income in Developing Countries*. USDA, ERS/FAS Economic Report 23. Washington, 1965.

Stryker, J. D. "Exports and Growth in the Ivory Coast: Timber, Cocoa, and Coffee." In Pearson and Cownie, 1974.

Stutley, Peter. "Government Intervention in Agricultural Marketing." In Hunter et al., 1976.

Thodey, A. R. "Marketing of Staple Food in Western Nigeria." 3 vols. Stanford Research Institute. Menlo Park, Calif., March 1968.

———. "Analysis of Staple Food Price Behavior in Western Nigeria." Ph.D. dissertation, University of Illinois, 1969.

Thompson, Virginia, and Adloff, Richard. *French West Africa*. Stanford, 1958.

Thrower, N. J. W., ed. Map Supplement No. 6 in *Annals of the Association of American Geographers*. Washington, March 1966.

Timmer, C. P. "On Measuring Technical Efficiency." *Food Research Institute Studies* 9 (1970):99–171.

Todaro, M. P. "Income Expectations, Rural-Urban Migration and Employment in Africa." *International Labour Review* 104 (1971):387–413.

Trapnell, G. G., and Clothier, J. N. *The Soils, Vegetation and Agricultural Systems of North Western Rhodesia*. Lusaka, 1937.

Uchendu, V. C., *The Igbo of Southeast Nigeria*. New York, 1965.

———. "Socioeconomic and Cultural Determinants of Rural Change in East and West Africa." *Food Research Institute Studies* 8 (1968):225–42.

———. "Field Study of Agricultural Change: The Cocoa Farmers of Akim-Abuakwa: Eastern Region, Ghana." Food Research Institute Study of Economic, Cultural, and Technical Determinants of Agricultural Change in Tropical Africa, Preliminary Report No. 8. Stanford, 1969, processed.

———. "The Extended Family and Employment." In *Conference of Directors of Social and Economic Research and Training Institutes*. Organisation for Economic Co-operation and Development (OECD). Paris, 1970a.

———. "The Impact of Changing Agricultural Technology on African Land Tenure." *Journal of Developing Areas* 4 (1970b):477–86.

Uchendu, V. C., and Anthony, K. R. M. "Field Study of Agricultural Change: Bawku District, Ghana." Food Research Institute Study of Economic, Cultural, and Technical Determinants of Agricultural Change in Tropical Africa, Preliminary Report No. 7. Stanford, 1969.

———. *Agricultural Change in Teso District, Uganda*. Nairobi, 1975a.

———. *Agricultural Change in Kisii District, Kenya*. Nairobi, 1975b.

Uganda, Ministry of Agriculture and Co-operatives. *Report on Uganda Census of Agriculture*, Vol. 1. Entebbe, 1965.

United Kingdom, Central Office of Information. *Agriculture in the United Kingdom Dependencies*. London, 1961.

United Nations. *Yearbook of International Trade Statistics 1975*, Vol. 1. New York, 1976a.

———. *Demographic Yearbook 1975*. New York, 1976b.

———, Department of Economic and Social Affairs. *World Economic Survey, 1975: Fluctuations and Development in the World Economy*. New York, 1976c.

———, Secretariat. "Estimates of Crude Birth Rates, Crude Death Rates, and Expectations of Life at Birth, Regions and Countries, 1950–1965." Population Division, Department of Economic and Social Affairs, ESA/P/WP/38. New York, February 22, 1971.

United States, Department of Agriculture (USDA). *The Income Situation.* Washington, July 1963.

———/Agency for International Development (USDA/AID). *Rice in West Africa.* Washington, 1968.

Vail, D. J., "Induced Farm Innovation and Derived Scientific Research Strategy: The Choice of Techniques in Developing Smallholder Agriculture in Land Abundant Areas." *Journal of Rural Development* 6 (1973):1–17.

Verger, Pierre. *"Influence du Brésil au Golfe du Bénin,"* In *Les Afro-Américains, Mémoires de l'Institut Français d'Afrique Noire* No. 27. Dakar, 1953.

Webster, C. C., and Wilson, P. N. *Agriculture in the Tropics.* London, 1966.

Welsch, D. E. "The Rice Industry in the Abakaliki Area of Eastern Nigeria, 1964." Ph.D. dissertation, Michigan State University, 1964.

Westley, Sidney, and Johnston, B. F., eds. "Proceedings of a Workshop on Farm Equipment Innovations for Agricultural Development and Rural Industrialization." University of Nairobi, Institute for Development Studies, Occasional Paper No. 16. Nairobi, 1975.

Wharton, C. R., Jr., ed. *Subsistence Agriculture and Economic Development.* Chicago, 1969.

Whetham, E. H. *Agricultural Marketing in Africa.* London, 1972.

Whitney, Anita. "Marketing of Staple Foods in Eastern Nigeria." Michigan State University, Department of Agricultural Economics Report 114. East Lansing, September 1968.

Wilson, Charles. "Economy and Society in Late Victorian Britain." *Economic History Review* 18, 2d Series (1965):183–98.

Wilson, P. N., and Watson, J. M. "Two Surveys of Kasilang Erony, Teso, 1937 and 1953." *Uganda Journal* 20 (1956):182–97.

Wrigley, Gordon. *Tropical Agriculture: The Development of Production.* New York, 1969.

Zuckerman, P. S. "Different Smallholder Types and their Development Needs." *Journal of Agricultural Economics* 28 (1977):119–27.

# Name Index

# Subject Index

**Library of Congress Cataloging in Publication Data**

(For library cataloging purposes only)
Main entry under title:
Agricultural change in tropical Africa.

    Bibliography: p.
    Includes indexes.
    1.  Agriculture—Africa, Sub-Saharan.  2.  Agricul-
tural innovations—Africa, Sub-Saharan.  3.  Agriculture—Economic aspects—Africa, Sub-Saharan.
4.  Agricultural productivity—Africa, Sub-Saharan.
5.  Agricultural research—Africa, Sub-Saharan.
I.  Anthony, Kenneth R. M.
S472.A1A34      338.1'0967      78-58039
ISBN 0-8014-1159-9